Smart AD and DA Conversion

ANALOG CIRCUITS AND SIGNAL PROCESSING

Consulting Editor*: Mohammed Ismail. *Ohio State University

For other titles published in this series, go to
www.springer.com/series/7381

Pieter Harpe • Hans Hegt
Arthur van Roermund

Smart AD and DA Conversion

Dr.ir. Pieter Harpe
Holst Centre – IMEC
High Tech Campus 31
Eindhoven
5656 AE
The Netherlands
pieter.harpe@imec-nl.nl
web@pieterharpe.nl

Prof.dr.ir. Arthur van Roermund
Department of Electrical Engineering
Eindhoven University of Technology
Den Dolech 2
Eindhoven
5612 AZ
The Netherlands
A.H.M.v.Roermund@tue.nl

Dr.ir. Hans Hegt
Department of Electrical Engineering
Eindhoven University of Technology
Den Dolech 2
Eindhoven
5612 AZ
The Netherlands
J.A.Hegt@tue.nl

Series Editors

Mohammed Ismail
Dept. Electrical and Computer Engineering
Ohio State University
Neil Avenue
43210
Columbus
US
ismail@ee.eng.ohio-state.edu

Mohamad Sawan
Dépt. Génie Informatique
Ecole Polytechnique de Montreal
H3C 3A7
Montreal
CA
Mohamad.sawan@polymtl.ca

ISBN 978-90-481-9041-6 e-ISBN 978-90-481-9042-3
DOI 10.1007/978-90-481-9042-3
Springer Dordrecht Heidelberg London New York

Library of Congress Control Number: 2010927405

Cover design: eStudio Calamar S.L.

Printed on acid-free paper

Springer is part of Springer Science+Business Media (www.springer.com)

Contents

List of symbols and abbreviations ix

1. INTRODUCTION 1

2. AD AND DA CONVERSION 3

   1 Introduction 3

   2 Trends in applications 4

   3 Trends in technology 5

   4 Trends in system design 6

   5 Performance criteria 8

   6 Conclusion 9

3. SMART CONVERSION 11

   1 Introduction 11

   2 Smart concept 11

   3 Application of the smart concept 15

   4 Focus in this work 16

   5 Conclusion 17

4. SMART DA CONVERSION 19

   1 Introduction 19

   2 Area of current-steering DACs 21

   3 Correction of mismatch errors 24

   4 Sub-binary variable-radix DAC 25

5 Design example 33
6 Conclusion 40

5. DESIGN OF A SUB-BINARY VARIABLE-RADIX DAC 41
1 Schematic design 41
2 Layout 45
3 Self-measurement-circuit implementation 46
4 Experimental results 47
5 Conclusion 54

6. SMART AD CONVERSION 57
1 Introduction 57
2 Literature review 58
3 High-speed high-resolution AD conversion 60
4 Smart calibration 67
5 Conclusion 70

7. DESIGN OF AN OPEN-LOOP T&H CIRCUIT 73
1 Literature review 73
2 Design goal 75
3 T&H architecture 75
4 Sampling core architecture 76
5 Output buffer architecture 78
6 T&H design 88
7 Experimental results 91
8 Conclusion 102

8. T&H CALIBRATION 103
1 Introduction 103
2 T&H accuracy 104
3 T&H calibration method 105
4 Analog correction parameters 106
5 Digitally assisted analog correction 113
6 Simulation results 117
7 Implementation of the calibration method and layout 119
8 Experimental results 120
9 Conclusion 124

9. T&H CALIBRATION FOR TIME-INTERLEAVED ADCS 125
1 Introduction 125
2 Channel matching in time-interleaved T&H's 128
3 Channel mismatch calibration 129
4 Channel mismatch detection 132
5 Channel mismatch correction 145
6 Simulation results 147
7 Implementation of the calibration method and layout 149
8 Experimental results 150
9 Conclusion 153

10. CONCLUSIONS 155

References 157

Index 163

List of symbols and abbreviations

Symbol	Description	Unit
ADC	Analog to digital converter	
CSA	Current source array	
D_a	$R_{1,a} - R_{1,1}$	
DAC	Digital to analog converter	
DNL	Differential non-linearity	LSB
Δ	Time-skew error	s
ENOB	Effective number of bits	bit
ERBW	Effective resolution bandwidth	Hz
FoM	Figure of merit	J/conversion-step
FS	Digital full scale amplitude	
f_s	Sampling frequency	Hz
G_e	Gain error	
INL	Integral non-linearity	LSB
LSB	Least significant bit	
M	MLS order	
m	MLS length	
MLS	Maximum length sequence	
MSB	Most significant bit	
N	Resolution	bit
O_e	Offset error	V, LSB
$R_{a,b}$	Discrete cross-correlation of a and b	
R_s	Discrete auto-correlation of $s[n]$	
$r[n]$	Single-bit MLS	
$s[n]$	Multi-bit MLS	
SFDR	Spurious free dynamic range	
SNDR	Signal to noise and distortion ratio	
SNR	Signal to noise ratio	
THD	Total harmonic distortion	
T&H	Track and hold	
$u[n]$	Response to a multi-bit MLS	
V_{CM}	Common-mode voltage	V
V_{fs}	Analog full scale amplitude	V

Chapter 1

INTRODUCTION

The history of the application of semiconductors for controlling currents goes back all the way to 1926, in which Julius Lilienfeld filed a patent for a "Method and apparatus for controlling electric currents" [1], which is considered the first work on metal/semiconductor field-effect transistors. More well-known is the work of William Shockley, John Bardeen and Walter Brattain in the 1940s [2, 3], after which the development of semiconductor devices commenced. In 1958, independent work from Jack Kilby and Robert Noyce led to the invention of integrated circuits. A few milestones in IC design are the first monolithic operational amplifier in 1963 (Fairchild μA702, Bob Widlar) and the first one-chip 4-bit microprocessor in 1971 (Intel 4004).

Ever since the start of the semiconductor history, integration plays an important role: starting from single devices, ICs with basic functions were developed (e.g. opamps, logic gates), followed by ICs that integrate larger parts of a system (e.g. microprocessors, radio tuners, audio amplifiers). Following this trend of system integration, this eventually leads to the integration of analog and digital components in one chip, resulting in mixed-signal ICs: digital components are required because signal processing is preferably done in the digital domain; analog components are required because physical signals are analog by nature. Mixed-signal ICs are already widespread in many applications (e.g. audio, video); for the future, it is expected that this trend will continue, leading to a larger scale of integration.

Given the trend of mixed-signal integration, this leads to both new challenges and new opportunities with respect to the integrated analog components. Challenges are for example testing of the performance of analog components that are embedded inside a large system, or the fact that the IC technology is optimized for digital circuitry, which can be disadvantageous for analog components. On the other hand, the mixed-signal integration also gives

P. Harpe et al., *Smart AD and DA Conversion*,
Analog Circuits and Signal Processing, DOI 10.1007/978-90-481-9042-3_1,

opportunities, like the possibility to shift parts of the system from the analog to the digital domain, or vice versa. From this point of view, the aim of this work is to investigate concepts to improve the performance of analog components by making use of the opportunities that are offered by mixed-signal system integration. This 'smart' concept will be applied to analog-to-digital and digital-to-analog converters, as these components are essential in mixed-signal systems.

The outline of this book is as follows: Chapter 2 studies trends and expectations in converter design with respect to applications, technology evolution and system design. Problems and opportunities are identified, and an overview of performance criteria is given. In Chapter 3, the smart concept is introduced that takes advantage of the expected opportunities (described in Chapter 2) in order to solve the anticipated problems. Chapters 4 and 5 apply the smart concept to digital-to-analog converters. In the discussed example, the concept is applied to reduce the area of the analog core of a current-steering DAC. In Chapter 4, the theory is presented while Chapter 5 discusses the implementation and experimental results. Chapter 6 up to Chapter 9 focus on the application of the smart concept to analog-to-digital converters. The main goal here is to improve the performance in terms of speed/power/accuracy. Chapter 6 introduces the general concept and defines key factors for the analog design and the smart approach in order to achieve the targeted high performance. Then, Chapter 7 deals with the analog design of an open-loop track-and-hold circuit. Experimental results are presented and compared against prior art. In Chapters 8 and 9, two calibration techniques are presented and experimentally verified by using the track-and-hold from Chapter 7. Finally, conclusions are drawn in Chapter 10.

Chapter 2

AD AND DA CONVERSION

This chapter studies trends and expectations in converter design with respect to applications, technology evolution and system design. Problems and opportunities are identified, and an overview of performance criteria is given. In Chapter 3, the smart concept is introduced that takes advantage of the expected opportunities in order to solve the anticipated problems.

1. Introduction

Electronic systems perform functions on many different types of signals, like: audio, video, medical images or RF communication signals. Despite the large variety, all signals are analog by nature in the physical world. However, at present most of the signal processing or signal storage is preferably performed in the digital domain. This leads to the need for analog-to-digital and digital-to-analog conversion. Actual systems can include both AD and DA conversion, or either one of the two. A general view on AD conversion is given in Fig. 2.1: the analog input signal is transferred to the digital domain through an ADC. Dependent on the situation, analog signal processing can be applied before the actual conversion, like pre-amplification, filtering or demodulation. Also, digital signal processing can be applied after the conversion, like error correction, filtering or data compression.

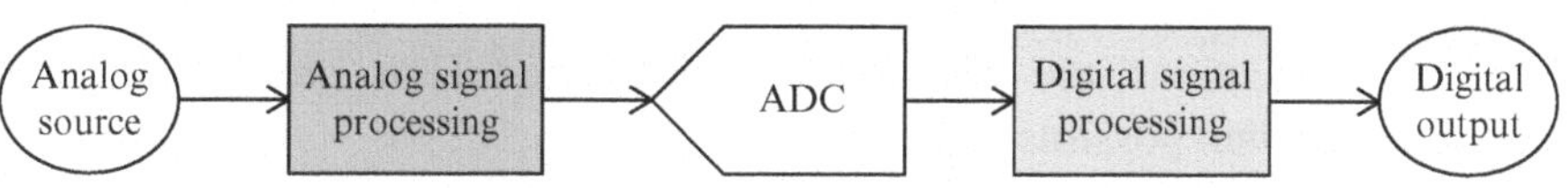

Figure 2.1. General view on AD conversion.

P. Harpe et al., *Smart AD and DA Conversion*,
Analog Circuits and Signal Processing, DOI 10.1007/978-90-481-9042-3_2,

A general view on DA conversion is given in Fig. 2.2: the digital input signal is transferred to the analog domain through a DAC. Dependent on the situation, digital signal processing can be applied before the actual conversion, like encoding or filtering. Also, analog signal processing can be applied after the conversion, like modulation or filtering.

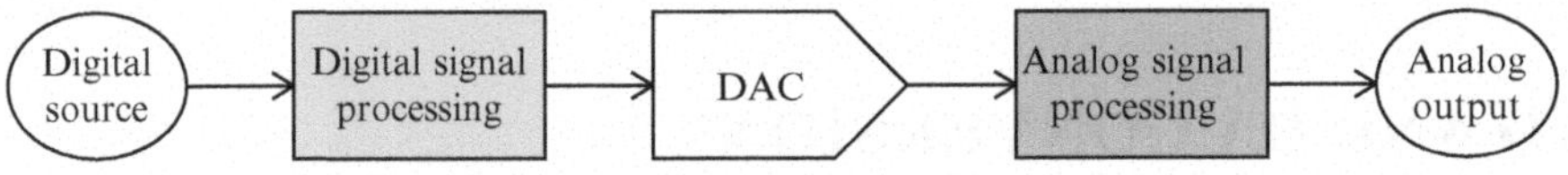

Figure 2.2. General view on DA conversion.

This work focusses on the actual AD and DA conversion, neglecting the other components of the signal processing chain.

2. Trends in applications

A trend in applications is that they typically demand a higher performance in terms of speed, accuracy, power consumption and chip-area. A motivation for the increase in demand is illustrated below.

- **Speed and accuracy**
  Over time, speed and accuracy requirements of an application are typically increasing. For example, the audio CD standard used 16 bit/44.1 kHz data, while current DVD players often use 24 bit/192 kHz. For digital still cameras, a 3 megapixel sensor with 12 bit dynamic range was state-of-the-art in 1999. Ten years later, state-of-the-art evolved to 25 megapixel sensors with 14 bit dynamic range. For wireless communication, the data-rate requirements are increasing, thus leading to a higher speed and/or accuracy requirement for the converters.

- **Power**
  In many situations, the power consumption is an important factor: e.g. to prevent problems due to thermal heating or to extend the lifetime of a battery-operated device. For example, for the previously mentioned still cameras, the battery lifetime improved from 400 shots to 4,000 shots on one battery. Though the reasons for this improvement are diverse, it shows that reducing power consumption is an important feature.

- **Area**
  For all applications, the chip-area is a cost-factor. Thus, area reduction is preferable when possible.

Concluding, the trends in applications lead to the challenge that more and more performance is expected from the AD/DA converters. To meet this challenge, two important opportunities are: trends in technology and trends in system design. Both these factors might be used beneficially as will be discussed in the following sections.

3. Trends in technology

In technology development, the most important trend is down-scaling of the devices for each technology generation. Next to that, also the power supply is being reduced. Figure 2.3 shows the development of technology and power supply as a function of time for the UMC foundry [4]. While the technology scales down at a relatively constant rate, the power supply scaling flattens out because of threshold limitations.

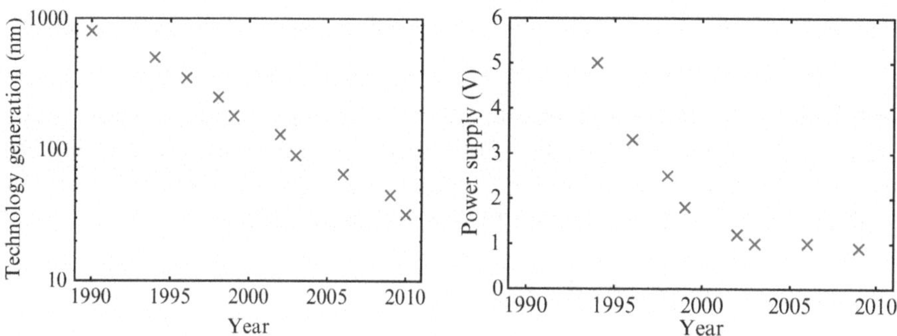

Figure 2.3. Technology scaling and supply scaling as a function of time.

For digital designs, both trends are beneficial as they enable:

- **Higher speed**
  Because of the reduction of the dimensions of the devices and the interconnect, the associated capacitances are also reduced, resulting in a higher speed of operation.

- **Lower power-consumption**
  The power consumption of digital circuits is proportional to CV_{DD}^2. As both the capacitances and the supply are being reduced, the power consumption will also decrease.

- **Smaller area**
  The smaller dimensions result in a smaller area or a higher integration density.

On the other hand, the same trends are not necessarily beneficial for analog designs. For a noise-limited system, the noise-power requirement (kT/C) results in a minimum C-value to be implemented. Moreover, as the supply scales down, the signal power most likely also scales down, thus requiring even an increase in C to maintain the SNR. Because of that, analog circuits do not directly take advantage of scaling to achieve a higher speed or a lower power-consumption. Apart from that, technology scaling also complicates analog design because of the following reasons:

- **Short channel effects**
  For smaller transistor geometries, secondary effects become more and more important. Because of that, the complexity of the transistor behavior increases, which complicates accurate circuit design.

- **Low voltage operation**
  The reduced supply voltage limits the number of transistors that can be stacked, which complicates the implementation of certain circuit topologies.

While digital circuit design benefits from technology scaling, analog circuit design is getting more complicated. For mixed-signal designs, like AD and DA converters, it seems a logical option to shift some of the analog problems to the digital domain to be able to benefit from technology scaling.

4. Trends in system design

As mentioned previously in Chapter 1, there is a tendency to integrate more and more components of a system into a single chip. As signal processing is predominantly performed in the digital domain while physical signals are analog by nature, this leads to the integration of analog and digital components in one chip, resulting in mixed-signal ICs. For the integrated analog components, the system integration offers both new challenges and new opportunities. Some of the challenges that become more important because of the mixed-signal integration are the following:

- **Hostile environment**
  Digital circuits create a hostile environment for the analog circuits by causing interference, which potentially reduces the performance of the analog circuits.

- **Testing**
  A stand-alone analog component can be tested directly for functionality or performance. On the other hand, analog components embedded in a large integrated system cannot be accessed directly, thus complicating test methodologies. A dedicated test-mode or an internal self-test strategy might be necessary to facilitate testing.

- **Yield**
  Especially when combining many different components into one integrated chip, the overall yield might be affected adversely by critical components.

Also, when a single component has an unsatisfactory performance, the total system might fail. In general, design for high yield becomes more important for integrated components when compared to stand-alone components.

- **Technology portability**
Though technology portability is a general issue, it becomes more relevant for mixed-signal ICs. For stand-alone analog components, a suitable technology might be selected by the designer. However, for integrated mixed-signal designs, the technology is most likely determined by the digital part and the analog circuits have to adopt this technology. Thus, analog designs are required that perform well in digital technologies. Moreover, they should be portable to future technologies as well, as the digital part of the system benefits from scaling. A second issue with respect to portability is the fact that expertise on analog design is required to transfer an existing design to a new technology.

- **Flexibility**
Many digital systems are flexible by using programmable hardware like microprocessors or FPGAs. Flexibility in mixed-signal integrated systems can be a useful aspect as it widens the application range for a single design and it allows software updates to accommodate post-production modifications. To achieve a higher level of flexibility, also the analog components need to be flexible (e.g. speed, power, accuracy), while maintaining an appropriate performance level.

- **Design time and risk**
The digital design flow is highly automated, which reduces the design time and risk, especially when porting an existing design to a new technology. On the other hand, analog circuit design, simulation and layout are mainly manual tasks with a potentially longer design time and a higher risk. Even when porting an existing design to a new technology, schematics, simulations and layout need to be redone to a large extent. Techniques to reduce the design time and risk of analog components are required to suit better to the digital design flow, and to prevent that analog design becomes the overall design bottleneck.

Apart from these challenges, mixed-signal integration also gives opportunities for analog circuit design:

- **Reuse of resources**
In a large integrated system, not all components will be actively used at all times. Especially the digital hardware might have free time-slots in which it can be used for other purposes like self-test or calibration of analog components. The freedom to use these already available resources can enhance

the overall system performance. Particularly in case of flexible digital platforms like FPGAs or microprocessors, hardware reuse can be implemented in a relatively simple way.

- **System optimization**
  In integrated systems, the main goal is not to optimize the individual components, but to optimize the overall system. Then, the components can be optimized given the specific task inside the larger system. Taking this approach into account, a better optimization might be possible when compared to standard stand-alone circuit design. Also, in a system-level approach, problems can be shifted from the analog to the digital domain whenever this is beneficial for the overall performance.

5. Performance criteria

From the described trends and requirements with respect to applications, technology and system design, a set of relevant performance criteria for AD and DA converters can be identified. Classified into these three trends, the following list of performance criteria is obtained:

- **Application driven criteria**
  - Speed
  - Accuracy
  - Power consumption
  - Area
- **Technology driven criteria**
  - Short-channel effect compatibility
  - Low-voltage compatibility
- **System driven criteria**
  - Interference compatibility
  - Testability
  - Yield
  - Portability
  - Flexibility
  - Design time and risk

Typically, the application driven performance criteria are the most important ones; these are also the most widely used performance criteria for the evaluation of converters. However, in view of technology development and system integration, it will become necessary to achieve sufficient performance for the other criteria as well.

6. Conclusion

In this chapter, it was shown that trends in applications, technology and system design complicate analog circuit design. It was also shown that apart from these challenges, technology scaling and system integration offer new opportunities to improve the performance of analog circuits. In the next chapter, a concept is introduced that takes advantage of these opportunities in order to solve the anticipated problems.

Chapter 3

SMART CONVERSION

This chapter introduces the smart concept for AD/DA converters, that aims at improving performance in a way that suits to the trends in technology and system design as described in Chapter 2. First, the smart concept (as published in [5]) will be defined. Then, various applications of the concept will be discussed and the main focus of this work will be explained.

1. Introduction

In the previous chapter, several challenges in AD/DA converter design were identified, namely:

- The performance in terms of speed, accuracy, power-consumption and area should improve because of the increasing demand from the applications.
- Technology properties limit the achievable performance. Moreover, the achievable performance range might be reduced for future technologies because of process down-scaling.
- Mixed-signal system integration introduces new challenges with respect to e.g. testing, yield, portability, interference.

Especially the first two trends are contradictory: the performance demand increases while for some situations, the intrinsically achievable performance decreases. Because of that, a solution is required that can overcome the intrinsic limitations.

2. Smart concept

The smart converter concept[1] implies on-chip intelligence to extract information after production in order to improve the performance beyond intrinsic limitations. The concept includes three main components:

P. Harpe et al., *Smart AD and DA Conversion*,
Analog Circuits and Signal Processing, DOI 10.1007/978-90-481-9042-3_3,

1. **Extraction** of information: to obtain more knowledge about e.g. imperfections that limit the performance, or specific requirements from the application.

2. **Processing** of this information: to decide how the performance can be optimized.

3. **Correction**: to realize the performance optimization.

The fundamental reason why a smart converter can potentially achieve a better performance than a conventional converter is because it has more information available. When appropriately used, this additional knowledge should allow a better optimization of the performance. Note that a-priori expertise is still required to decide how to extract, process and correct appropriately.

A second motivation for the smart concept is that several parts (as will be shown later) can be implemented in the digital domain. By doing so, these parts will benefit automatically from technology scaling, leading to a future-proof solution.

An illustration of the smart concept is shown in Fig. 3.1; the various options for extraction, processing and correction will be discussed in the following sections.

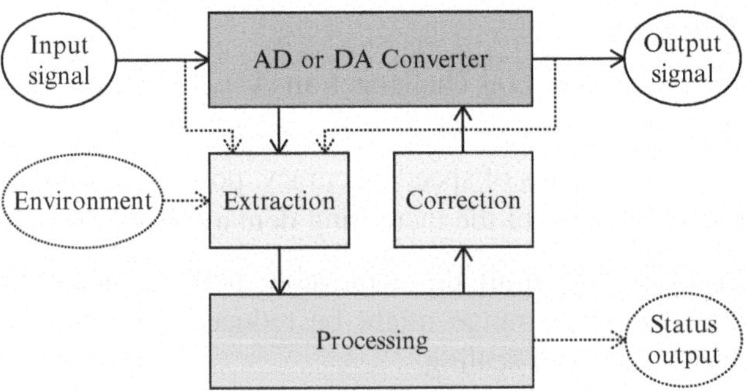

Figure 3.1. Smart converter concept.

2.1 Extraction

The first step of the smart concept is the extraction of performance-relevant information. Three different information sources can be distinguished: system information, signal information and environmental information:

- **System information**
  In this case, that is information from the ADC or DAC, like: functionality, mismatch of components (random or due to process-spread), non-linearity, frequency dependent behavior. As an example, consider the mismatch of

components: before production only the statistics of the mismatch are known. After production, for a specific chip, the mismatch has a deterministic value. When this value can be measured on-chip it gives more precise knowledge on the mismatch compared to the pre-production statistical information. With this additional information and an appropriate correction technique, this imperfection could be counteracted to improve the performance.

- **Signal information**
  Information from signals at the input, output or at an intermediate stage, like: amplitude, bandwidth, probability density function, spectral properties. For example, when the input signal has a limited amplitude, the converter's resources could be optimized for that specific amplitude instead of being optimized for the full-scale range of the converter.
- **Environmental information**
  Environmental information includes information from the ambience (e.g. temperature, supply voltage), the user and the application. For example, dependent on the user/application requirements, the relevant performance criteria might change, and thus require a different optimization goal of the smart converter.

When considering mixed-signal circuits like AD and DA converters, part of the information to be extracted will be available in the digital domain and part of it will be available in the analog domain. In both cases, additional hardware might be required to extract the information.

2.2 Processing

The second step of the smart concept is to process the extracted information to optimize the performance. Two parts can be distinguished in the processing block: a first part in which the relevant information is extracted from the raw data and a second part where the information is processed to achieve a suitable correction. Whether these parts are necessary depends on the methods used for extraction and correction:

- **Processing of extracted information**
  Dependent on the extraction method, the required information for the performance optimization can be directly available, or it can be embedded in another signal. For example, consider the determination of the offset of an ADC. As a first option, one could set the ADC-input to zero and measure the digital output code, which then directly corresponds to the offset, and no additional processing is required to extract the relevant information. Alternatively, under the assumption that the applied (unknown) input signal has no DC component, the average output code of the ADC corresponds to

the offset. In this case, processing (namely averaging) is required to obtain the offset information from the overall signal.

- **Optimization algorithm**
  Dependent on the correction method, an optimization algorithm might be required to achieve the optimal performance. Considering the correction of the offset of an ADC, a first option could be to apply a digital correction by subtracting the measured offset digitally for each conversion. In that case, no optimization algorithm is required as the measured offset can be applied directly in the correction method. In a second case, the offset might be corrected by calibration of the analog reference voltage. In that case, several iterations might be necessary to find the best possible setting of the analog reference voltage to minimize the offset.

From the above examples, it can be understood that the complexity of the processing algorithm is strongly dependent on the methods used for extraction and correction.

As shown in Fig. 3.1, the processing algorithm could also include a 'status output', which gives relevant information about the status of the converter to the outside world, for example to facilitate an on-chip self-test for specific performance parameters. For example, suppose a smart correction algorithm is used to compensate a certain imperfection. When the algorithm runs out of range, it might suggest that the converter does not meet the target specification.

The processing algorithm can be implemented either in the analog or in the digital domain. A digital solution seems the most logical choice for the following reasons:

- Digital hardware offers flexibility and memory while it does not add unknown imperfections.

- Given the technology trends described in the previous chapter, digital processing will become cheaper for each technology generation. Thus, a digital implementation can benefit automatically from technology scaling.

- Given the context of mixed-signal integrated systems, it is expected that a large amount of programmable digital resources is present in the system. During free time-slots (e.g. at startup of the system), this hardware could be temporarily used to perform the processing algorithm.

2.3 Correction

The third step of the smart concept is to perform the actual correction to optimize the performance. Several examples of correction methods are the following:

- **Digital correction**
  Digital signal processing can be used to counteract measured imperfections. In case of ADCs, this results in digital post-correction; for DACs, this results in digital pre-correction. For example, offset could be compensated digitally for both ADCs and DACs by subtracting the measured offset in the digital domain.

- **Analog correction**
  By tuning analog components, specific imperfections can be corrected. For example, the unit elements inside a converter suffer from mismatch. This could be corrected by adding calibration elements to fine-tune the elements to their optimal value.

- **Mapping**
  A mapping method optimizes the overall performance by selecting a certain order or a certain combination within the available resources. For example, the mismatch of the unit elements inside a converter results in a limited performance. By reordering the unit elements, the overall performance can be optimized, even though the individual errors remain the same.

3. Application of the smart concept

The smart concept is a general idea to obtain knowledge on-chip in order to enhance the performance. As there are many different performance limitations and performance criteria, the smart concept can be applied in many different ways. The overview below illustrates how the smart concept can be applied advantageously for each of the performance criteria, defined in Section 5.

- **Speed, accuracy, power consumption, area**
  There are many trade-offs between these performance criteria. To illustrate the possibilities of the smart concept, mismatch of components is considered. In a conventional converter, the accuracy is directly related to the mismatch of the unit elements. For sufficient accuracy, large elements are required, resulting in a large area and increased parasitics, which can cause speed limitations and an increase in power consumption. When the mismatch errors could be determined and corrected on-chip, much smaller elements could be used, thereby giving potential improvements in accuracy, area, speed and power consumption.

- **Compatibility to future technologies**
  Efficient low-voltage compatible circuits (e.g. open-loop instead of closed-loop amplifiers) often do not meet the performance requirements (e.g. because of non-linearity). The smart concept allows the use of these circuits as their associated limitations can now be overcome by a smart correction.

- **Interference compatibility**
  On-chip sensors could be implemented to measure the interference. Based on that, a mapping method for both analog and digital hardware could be used to reduce the interference for the most critical analog blocks.

- **Testability**
  The extracted information in a smart converter can be used to enable on-chip self-test by measuring relevant test parameters like functionality or accuracy.

- **Yield**
  Instead of relying on intrinsic performance, additional correction resources can be built into a smart converter to allow compensation of a larger performance spread, which can improve the yield.

- **Portability, design time and risk**
  In an intrinsic design, the performance is determined by the technology, thereby complicating portability as the performance has to be verified again in the new technology. In a smart approach, the technology limitations can be overcome. Because of that, the technology has less influence on the performance which simplifies portability and reduces design time and risk.

- **Flexibility**
  As a large part of the smart circuitry can be implemented with programmable logic, the smart concept allows optimization of a specific converter for various requirements, thereby adding flexibility to the design.

While the list illustrates the versatility of the smart concept, it should not be expected that all smart designs improve all these parameters at once. For example, in [6] a smart redesign of a conventional pipelined ADC [7] is proposed: an open-loop amplifier is used instead of a closed-loop amplifier to reduce the power consumption. Then, the non-linearity of the open-loop structure is corrected by a smart approach. Nonetheless, the FoM of the smart converter (2.3 pJ/conv.step) is worse than the FoM of the original design (1.3 pJ/conv.step).[2] However, the smart design performs better with respect to technology compatibility, which makes it an attractive solution for future technologies.

4. Focus in this work

The aim of this work is to investigate the feasibility of relevant smart AD and DA converter concepts. The main focus will be on the application driven performance criteria (speed, accuracy, power consumption and area), while the technology compatibility is also taken into account. In Chapter 4, a smart concept for DA converters is proposed to improve the performance with respect to area; the chip implementation of this concept will be discussed in Chapter 5.

In Chapter 6, a smart concept for AD converters is proposed to improve the performance with respect to the speed/power/accuracy trade-off; the chip implementation is shown in Chapters 7 up to 9.

5. Conclusion

In this chapter, the smart converter concept was proposed. It was shown that this concept can overcome some important technology limitations, thereby improving the performance and the compatibility to future technologies. As the concept is implemented on-chip, it is also compatible with mixed-signal integrated systems. Because of these reasons, the smart concept suits to the trends in applications, technology and system design.

Notes

1. In this work, the smart concept is limited to AD and DA converters. However, the smart concept can be applied in a similar way to other circuits or systems.
2. The FoM definition is given in Chapter 6.

Chapter 4

SMART DA CONVERSION

This chapter applies the previously presented smart concept to digital-to-analog converters. In the proposed scenario, the smart concept will be used to reduce the chip-area of a current-steering DAC while maintaining overall accuracy. In this chapter, the theory of the approach will be studied, while the actual proof-of-concept by means of a chip implementation will be given in Chapter 5. Parts of this chapter have been published previously in [8–10].

1. Introduction

In the previous chapter, the versatility of the smart concept was explained. Comprehensibly, in reality one can only demonstrate the feasibility of a limited part of the overall concept. In this chapter, one relevant item from the smart concept will be selected for further investigation and actual chip implementation. In Chapter 6, where the smart concept will be applied to AD converters, different items will be chosen to show another view of the smart concept.

Existing work that includes some of the aspects of the smart concept aims for improving performance by means of correction, calibration or mapping. Typically, the most important goal is to improve the accuracy by counteracting the effect of a certain error (or a set of errors). An overview of error mechanisms and correction methods is described in [11]; the enumeration below gives a summary of the described error mechanisms and examples of work that counteract the errors by means of correction.

- Amplitude mismatch, e.g. [12, 13]
- Timing mismatch, e.g. [14]
- Harmonic distortion, e.g. [11]
- Data-switching errors, e.g. [15]

P. Harpe et al., *Smart AD and DA Conversion*,
Analog Circuits and Signal Processing, DOI 10.1007/978-90-481-9042-3_4,

Apart from improving accuracy, the methods can also have other benefits. For example, when timing mismatch can be corrected, it does not only improve accuracy, it can also extend the usable frequency range, thus increase the speed of operation. Or, when amplitude mismatch can be corrected, one might use smaller (and thus less accurate) elements to reduce the area, and still meet the accuracy requirements.

In this work, the smart concept will be applied with as main goal to reduce the area of the analog part of a DAC as much as possible. Because of the area reduction, a large amount of amplitude mismatch can be expected. Thus, a digitally-implemented smart solution will be applied to maintain high accuracy. The motivation for this goal is that currently, DAC designs do not scale down in size as fast as technology-scaling would permit, because of accuracy requirements. At the same time, digital components do scale down with technology. As a consequence, in mixed analog/digital systems, the DAC becomes relatively larger in size compared to the digital hardware. Because of that, it is worthwhile to investigate solutions to reduce the area of the DAC.

When the area of the DAC is considered, one can refer to either the overall DAC area or the analog-core area:

- **Overall DAC area**
  Complete area of the DAC (analog and digital parts) and area of the add-on digital circuitry for the smart solution.

- **Analog core area**
  Area of the analog parts of the DAC only, excluding digital parts inside the DAC or digital parts added to the DAC.

In situations where the DAC is a stand-alone device, the most relevant goal would be to minimize the overall DAC area instead of minimizing the analog core area. However, the goal in this case is to minimize the analog core area, even when it comes at the cost of an increased area of the digital part. This is motivated by the following reasons:

- **Application in large-scale mixed-signal ICs**
  A first motivation is that the smart concept envisions large-scale mixed-signal integrated systems. In these systems, there are applications for which the analog core area is more important than the overall area; two examples where the analog core area is the most relevant area to be optimized are given here:
  A first example is when the DAC is used as an accurate on-chip test-signal generator during calibration of another on-chip component. After calibration, the digital hardware required for the DAC can be reprogrammed for another task. Then, because of the flexibility of the digital hardware, the only overhead in terms of area is coming from the analog DAC core.

Another example is a general-purpose digital chip in which, for a few applications, a DAC is required. As the chip is a general-purpose IC and the DAC is only used for a few applications, it would be expensive to implement a large-size DAC on all these devices. On the other hand, when the DAC area would be very small, the overhead would be acceptable to implement the DAC on all chips. Then, only for the few applications that actually use the DAC, digital resources have to be allocated to control the DAC. Even when a substantial amount of digital resources would be required, this can still be the most effective solution on the average, as most application do not use the DAC.

- **Different scaling for analog and digital**
  A second motivation, as explained before, is that digital scales down with technology much faster than analog. Thus, reducing the area of the analog core will also reduce the overall area on the long term.

Given this background, the aim of area reduction will be limited to the analog core area in this work.

This chapter starts with a discussion on area constraints of DA converters in Section 2. Existing approaches to reduce the area are reviewed in Section 3, and a new concept for area reduction is introduced in Section 4. A design example is shown in Section 5 and conclusions are drawn in Section 6.

2. Area of current-steering DACs

A simplified view of the basic architecture of a current-steering DAC is shown in Fig. 4.1. The digital input code is optionally translated in a decoder (e.g. a binary-to-thermometer decoder). Then, the decoded digital signal drives a set of switch drivers that control the analog switches. By means of the analog switches, the current sources from the current-source-array (CSA) are connected to either the positive or the negative output of the DAC, thereby generating the output signal as a function of the applied input code.

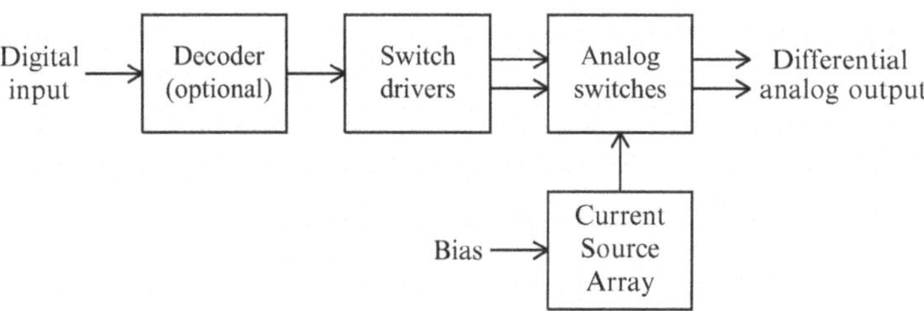

Figure 4.1. Architecture of a current-steering DAC.

The decoder, switch drivers and switches can be scaled down as much as the technology permits. However, the size of the CSA is limited by the required accuracy of the DAC. In Table 4.1, an overview of recent work on current-steering DACs is given. All these designs are based on intrinsic-accuracy, i.e. they do not employ special algorithms or calibration to enhance the performance. The achieved static accuracy is indicated by the $ENOB$ that was calculated as:

$$ENOB = resolution - log_2\Big(max(INL_{max}, DNL_{max})\Big) - 1, \tag{4.1}$$

where INL_{max} and DNL_{max} are expressed in LSB of the native resolution.[1] From the table, where both the total area of the DAC and the (estimated) area of the current-source-array (CSA) are given, it can be observed that:

- The area increases rapidly as a function of the resolution/accuracy.
- The CSA is a major component in the overall area.

Table 4.1. Area of intrinsic-design DA converters.

Reference	Technology (μm)	Sampling rate (MSps)	Resolution (bit)	ENOB (bit)	Total area (mm^2)	CSA area (mm^2)
[16], 2001	0.35	1,000	10	11.3	0.35	0.1
[17], 2006	0.18	250	10	12.3	0.35	0.2
[18], 2004	0.18	600	12	11.0	1.1	0.4
[19], 2004	0.18	320	12	12.3	0.44	0.2
[20], 2001	0.35	500	12	12.7	1.0	0.5
[21], 2009	0.065	2,900	12	12.0	0.3	0.1
[22], 2004	0.18	1,400	14	12.2	2.5	0.4

The reason why the area increases so fast as a function of the resolution, and why the CSA is a major contributor to the overall area is because of accuracy requirements: to achieve appropriate INL/DNL performance, the current sources in the CSA need to be matched to each other. The higher the resolution, the more stringent the matching becomes, which translates into a larger area. A lower-bound for the required area for the CSA will be derived next.

The target requirement is set such that a 3σ mismatch-error still achieves 0.5 LSB INL performance. From [23], it is known that for a resolution of N bit, this leads to the following matching constraint:

$$\frac{\sigma_u}{I_u} = \frac{1}{6\sqrt{2^{N-1}}}, \tag{4.2}$$

where σ_u/I_u represents the relative standard deviation of the unit current element (the LSB current source); i.e. σ_u is the absolute standard deviation and I_u the nominal current of the unit element.

From amongst others (a.o.) [11], the standard deviation of the unit element can be expressed in terms of the mismatch parameters of the technology (A_β and A_{V_t}), the biasing condition ($V_{gs} - V_t$) and the dimensions of the unit element (W_u and L_u):

$$\frac{\sigma_u}{I_u} = \sqrt{\frac{A_\beta^2 + 4A_{V_t}^2/(V_{gs} - V_t)^2}{2W_u L_u}} \tag{4.3}$$

For a given technology (e.g. a 0.18 µm CMOS technology: $A_\beta = 2\%$ µm and $A_{V_t} = 4\text{mV}\mu\text{m}$), and an estimate of the biasing condition (e.g. $V_{gs} - V_t = 0.5$ V), (4.3) simplifies to:

$$\frac{\sigma_u}{I_u} = \frac{0.018}{\sqrt{W_u L_u}} \tag{4.4}$$

Note that for newer technologies, A_β and A_{V_t} are typically improving slightly, thereby improving the matching. However, at the same time, new technologies allow less voltage-headroom, such that the loss of $V_{gs} - V_t$ counteracts the improved matching properties. As a consequence, little improvement of the relation given by (4.4) is expected for DACs designed in newer technologies.

Combining equations (4.2) and (4.4) yields a relation between the resolution and the area of the unit element:

$$W_u L_u = 0.006 \cdot 2^N \tag{4.5}$$

As the DAC is composed of a total of 2^N unit elements, the total gate-area becomes:

$$A_{dac} = 0.006 \cdot 4^N \ [\mu\text{m}^2] \tag{4.6}$$

Note that this equation reveals that the area increases with a factor of four for each additional bit of resolution; this explains why in Table 4.1 the area increases so rapidly as a function of the resolution. Table 4.2 shows the theoretical minimum area of the CSA for various resolutions, based on (4.6). With the used mismatch-parameters, the values are only valid for a specific 0.18 µm technology. But, as explained previously, it is expected that the relation is not affected strongly by the evolution of technology. In practice, the CSA area will become even larger due to overhead like wiring, source/drain/gate connections, spacing, etc. Especially for 12-bit resolution and higher, the required area becomes large. For these situations, an area-reduction technique becomes necessary if the area is a critical design parameter.

Table 4.2. Lower bound for the active area of the CSA in a 0.18 µm technology.

Number of bits (N)	8	10	12	14	16
Area (A_{dac}) mm^2	0.0004	0.006	0.1	1.6	26

3. Correction of mismatch errors

The fact that for high-resolution current-steering DACs an area reduction technique is advantageous has been recognized, and led to the development of correction techniques that maintain the final overall accuracy while starting with intrinsically less-accurate elements. By requiring less intrinsic accuracy, the area of the CSA can be reduced. A classification and an explanation of the existing methods for area reduction can be found in [11].

Table 4.3 gives an overview of recent work on designs that include some form of mismatch-correction to achieve area-reduction. In [24], a reshuffling method is used to order the elements in such a way that the accuracy improves. In the other approaches, the main current sources are calibrated by trimming or by adding a calibration current to compensate for the mismatch. In the table, the CSA area includes the area of the nominal current sources plus the area of the calibration sources.

Table 4.3. Area of corrected DA converters.

Reference	Technology (μm)	Sampling rate (MSps)	Resolution (bit)	ENOB (bit)	Total area (mm^2)	CSA area (mm^2)
[13], 2005	0.25	50	12	11.7	1.1	0.3
[25], 2008	0.18	100	12	13.3	0.8	0.2
[24], 2007	0.18	200	14	12.5	3	0.28
[26], 2004	0.18	200	14	13.6	1	0.5
[27], 2001	0.18	100	14	14	1	0.3
[28], 2003	0.13	100	14	14.2	0.1	0.05
[29], 2000	0.35	100	14	14.5	11.8	1.8
[12], 2003	0.25	400	16	16	2	0.8

Figure 4.2 shows the area of the CSA for the references from Tables 4.1 and 4.3 as a function of the achieved accuracy. Also shown is the theoretical intrinsic CSA area limit (4.6).

From the figure, it can be concluded that:

- The corrected designs achieve a higher area-accuracy performance when compared to the intrinsic designs.
- The intrinsic designs can never achieve an area below the intrinsic CSA-limit; the corrected designs can achieve an area around or below the CSA-limit.
- In most cases, the corrected designs aim for a higher resolution and a higher accuracy than the intrinsic designs.

Though the calibration methods can overcome the intrinsic CSA area limitation, the area improvement is practically limited as the number of current

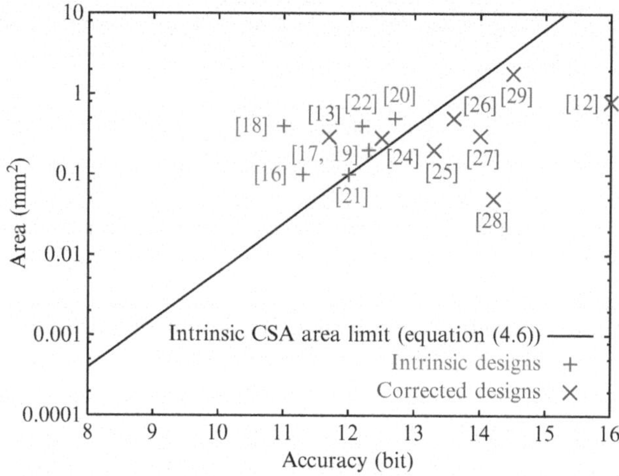

Figure 4.2. Area of the CSA for intrinsic and corrected designs.

elements increases to facilitate calibration: next to the main current sources, also calibration sources are required. As soon as the area of the unit elements becomes substantially reduced, it is not the *area* of the elements but simply the *amount* of elements that determines the overall CSA area. This is because the overhead caused by e.g. wiring and spacing will increase as a function of the number of elements, such that the overhead-area will become dominant.

Concluding from the results of intrinsic and corrected designs, there are two factors that determine the CSA area; and both should be minimized in order to minimize the CSA area:

- The area of the current-source elements.
- The number of current-source elements.

In the following section, an approach is proposed that aims at minimizing both these factors to achieve a further reduction of the CSA area.

4. Sub-binary variable-radix DAC

4.1 System overview

Pursuant to the aim of minimizing the area of the CSA and solving the related issues in the digital domain, a digitally pre-corrected DAC is proposed. A general view of the system is given in Fig. 4.3. The DAC core contains the switch drivers, switches and the area-minimized CSA. The mismatch-errors of the current sources are corrected by the digital pre-correction block that remaps the binary input codes to appropriate combinations of current sources. A built-in measurement algorithm is used to measure the actual deviations of

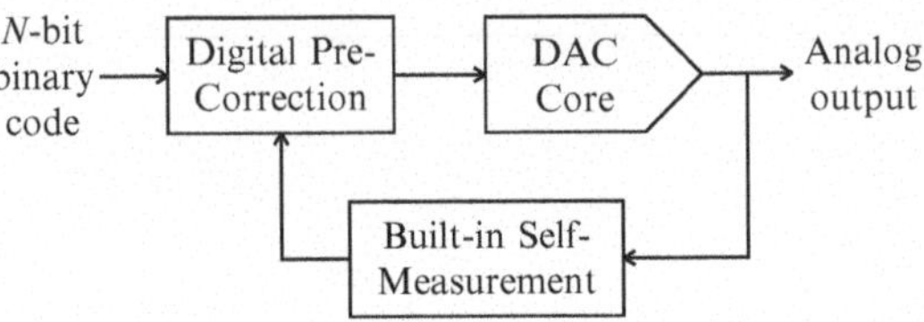

Figure 4.3. Digitally pre-corrected DAC with built-in self-measurement.

the individual current sources, such that the digital pre-correction algorithm can determine a suitable combination of current sources for each input code.

As a starting point for minimizing the CSA area, a first consideration is the segmentation of the current sources. The segmentation determines the number of independent current sources within the CSA. The three common options are:

- Binary architecture; composed of N binary-scaled sources.
- Thermometer architecture; composed of $2^N - 1$ unary-scaled sources.
- Segmented architecture; composed of a M binary-scaled and $2^{N-M} - 1$ unary-scaled sources, with $1 \leq M < N$.

From these options, the binary architecture has the least number of independent sources. As a consequence, according to the conclusion from the previous section, the binary architecture should have the potential to achieve the smallest CSA area.

However, when a binary architecture is used without any modification, digital pre-correction cannot correct all types of mismatch errors. Figure 4.4 illustrates several transfer curves for a 5-bit binary-scaled DAC: the nominal curve, the curve in case of a positive DNL error in the MSB, and the curve in case of a negative DNL error in the MSB. While considering the MSB only in

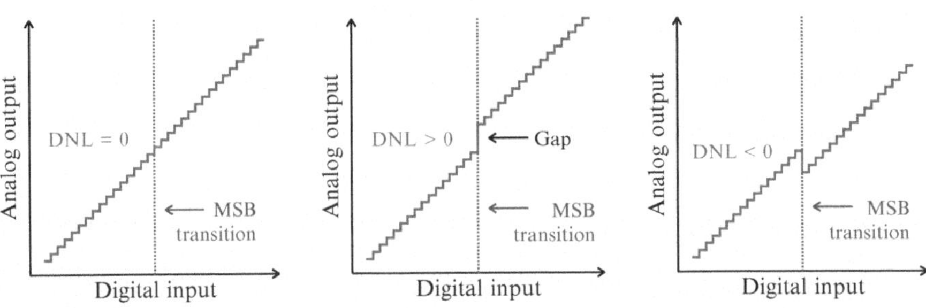

Figure 4.4. Transfer curve of a 5-bit DAC: no mismatch (left), a 'positive' (middle) and a 'negative' (right) deviation of the MSB.

this example, a comparable situation can occur with the other sources of the DAC. The large DNL error produced for DNL > 0 cannot be reduced with pre-correction, as pre-correction can only re-map the input code to an existing combination of current sources that approximates the desired output level. However, for DNL > 0, there is no combination of current sources available to fill the gap in the output range. On the other hand, the large DNL error for DNL < 0 can be corrected with pre-correction, as there is a 'gap-free' continuum of output levels. By digital re-mapping, the overlap (or non-monotonicity) of the curve can be removed to obtain a smooth transfer curve. However, as a side effect of the overlap, the full-scale range of this converter will be slightly smaller than usual.

In short, for the digital pre-correction to operate properly, 'gaps' (DNL > 0) are not allowed but 'overlap' (DNL < 0) is allowed. As the nominal transfer curve of a normal DAC is designed for DNL $= 0$, there is a 50% probability that a 'gap' will occur in reality. By means of *redundancy*, the probability of a 'gap' can be reduced to an arbitrary low value by design: instead of designing the nominal transfer curve as in Fig. 4.4 (left), it is designed as in Fig. 4.4 (right). Thus, redundancy introduces intentional overlap (DNL < 0) of the nominal transfer curve to guarantee that the continuum of the output range remains, also in case of mismatch. While the figure illustrates redundancy for the MSB only, in reality this redundancy requirement needs to be implemented for each bit of the converter. In a certain way, adding calibration current sources (as in a.o. [11]) adds redundancy. However, that approach also leads to a substantial increase of the amount of current sources; which is undesirable as explained in the previous section. Therefore, in this work, the use of a sub-binary radix is proposed to introduce redundancy while limiting the total number of current sources.

Note that the fundamental principle of a sub-binary radix for DACs is equivalent to the principle of a sub-binary radix for ADCs (as explained in e.g. [30]): in ADCs the redundancy is used to alleviate the effect of comparator mismatch, whereas in DACs, the redundancy is used to alleviate the effect of current source mismatch. Also note that, independently from this work, prior art on sub-binary radix DACs exists [31], but that work does not mathematically optimize the redundancy, uses a fixed instead of a variable radix, and requires a more complex measurement method.

In the following sections, the design of the DAC core with redundancy, the self-measurement structure and the digital pre-correction algorithm will be explained.

4.2 Redundancy

A normal N-bit binary converter is composed of $k = N$ current sources. These sources I_0 (LSB) up to I_{k-1} (MSB) are chosen relatively to the unit element I_u using the ratios α_0 up to α_{k-1}. The ratios α_i are chosen such that each source is exactly 1 LSB larger than the sum of all smaller sources:

$$\alpha_j = \sum_{i=0}^{j-1} \alpha_i + 1 \text{ for } 0 \leq j < k \,, \tag{4.7}$$

leading to the binary-scaled sequence of α's: 1, 2, 4, 8, 16, ... However, when due to mismatch one of the current sources is actually larger than expected, a 'gap' (as in Fig. 4.4) arises, that cannot be corrected with digital pre-correction. To avoid this situation, redundancy is added by making α_j intentionally smaller than the sum of all smaller sources plus one LSB:

$$\alpha_j < \sum_{i=0}^{j-1} \alpha_i + 1 \text{ for } 0 \leq j < k \tag{4.8}$$

An example of a sequence, fulfilling this constraint, is e.g.: 0.7, 1.3, 2.4, 4.6, 8.8, ... The amount of redundancy r_j for each source can be expressed as:

$$r_j = 1 - \alpha_j + \sum_{i=0}^{j-1} \alpha_i \text{ for } 0 \leq j < k \,, \tag{4.9}$$

Thus, for $r_j = 0$, there is no redundancy and (4.9) simplifies to (4.7). To maintain redundancy, for all sources the following requirement has to be satisfied:

$$r_j > 0 \text{ for } 0 \leq j < k \tag{4.10}$$

The more redundancy is added, the more severe deviations due to mismatch can be compensated by pre-correction. However, also note that the more redundancy, the more sources k have to be employed to compensate the full-scale reduction of the converter due to redundancy.

Due to the stochastic spread of the unit cells, the actual value of each source becomes a stochastic value $\underline{\alpha}_i$ with mean α_i (the designed value) and spread $\sqrt{\alpha_i}\frac{\sigma_u}{I_u}$. To guarantee that all required output levels can be produced with sufficient accuracy using pre-correction, the relations from (4.9) and (4.10), taking the stochastic spread of the sources into account, have to be fulfilled. This leads to the following set of requirements:

$$\underline{r}_j > 0 \text{ for } 0 \leq j < k \,, \text{ with:} \tag{4.11}$$

$$\mathbf{E}\{\underline{r}_j\} = 1 - \alpha_j + \sum_{i=0}^{j-1} \alpha_i$$

$$\sigma_{\underline{r}_j} = \frac{\sigma_u}{I_u}\sqrt{\sum_{i=0}^{j}\alpha_i}\,,$$

where $\mathbf{E}\{\underline{r}_j\}$ is the expectation of $\underline{r}_j$, thus the nominal built-in redundancy. $\sigma_{\underline{r}_j}$ is the spread of $\underline{r}_j$, which corresponds to the mismatch of the elements. When all constraints $\underline{r}_j$ are fulfilled, the largest 'gap' in the transfer curve is guaranteed to be less than 1 LSB. Thus, each required output level can be generated within ± 0.5 LSB by means of re-mapping the input code, which is sufficient for meeting the target accuracy. The desired probability of fulfilling each constraint $\underline{r}_j$ can be expressed as a desired level of confidence $\lambda\sigma$ with which the constraint has to be fulfilled:

$$P\{\underline{r}_j > 0\} = 1 - \frac{1}{2}\,\mathrm{erfc}\Big(\frac{\lambda}{\sqrt{2}}\Big) \tag{4.12}$$

The confidence level requires that:

$$\mathbf{E}\{\underline{r}_j\} - \lambda\sigma_{\underline{r}_j} = 0\,, \tag{4.13}$$

i.e., a $\lambda\sigma$ deviation from the nominal value $\mathbf{E}\{\cdot\}$ is still marginally acceptable for the target requirement (4.11). Using equation (4.11), (4.13) can be rewritten as:

$$\begin{aligned}
\mathbf{E}\{\underline{r}_j\} &= \lambda\sigma_{\underline{r}_j} \qquad (4.14)\\
1 - \alpha_j + \sum_{i=0}^{j-1}\alpha_i &= \lambda\frac{\sigma_u}{I_u}\sqrt{\sum_{i=0}^{j}\alpha_i}\\
\Big(1 + \sum_{i=0}^{j-1}\alpha_i\Big)^2 + \alpha_j^2 - 2\alpha_j\Big(1 + \sum_{i=0}^{j-1}\alpha_i\Big) &= \Big(\frac{\lambda\sigma_u}{I_u}\Big)^2\sum_{i=0}^{j}\alpha_i\\
&\Rightarrow\\
\alpha_j^2 - b_j\alpha_j + c_j &= 0, \text{ with}
\end{aligned}$$

$$\begin{aligned}
b_j &= 2\Big(1 + \sum_{i=0}^{j-1}\alpha_i\Big) + \Big(\frac{\lambda\sigma_u}{I_u}\Big)^2\\
c_j &= \Big(1 + \sum_{i=0}^{j-1}\alpha_i\Big)^2 - \Big(\frac{\lambda\sigma_u}{I_u}\Big)^2\sum_{i=0}^{j-1}\alpha_i
\end{aligned}$$

From this quadratic function of α_j, the values of α_j can be derived recursively given the relative spread of the unit cells and a desired confidence level:

$$\alpha_j = \frac{b_j - \sqrt{b_j^2 - 4c_j}}{2}, \qquad \text{with } b_j \text{ and } c_j \text{ as in (4.14).} \tag{4.15}$$

As opposed to previous work [31], where a fixed sub-binary radix was used, the presented approach utilizes a variable radix ρ_j. The radix is the ratio between two subsequent current sources:

$$\rho_j = \frac{\alpha_j}{\alpha_{j-1}} \tag{4.16}$$

The variable radix stems from the design procedure equalizing the error probability for each constraint (4.15). By equalizing the error probability and adapting the radix, instead of equalizing the radix and adapting the error probability as in [31], the same yield can be achieved with less redundancy and hence less current sources. Moreover, as in the presented approach, ρ_j approximates 2 for $j \to \infty$, the amount of sources k required for a converter with redundancy comes close to the minimal value N as used in a binary-weighted converter without redundancy. Simulation results to confirm the advantages of the variable-radix over the fixed-radix will be shown in Section 5.

4.3 Self-measurement

Before being able to pre-correct the mismatch errors of the current sources, a measurement procedure, measuring the actual values of the current sources, is required. After performing the self-measurement procedure at power-up, the actual values of the current sources are known in the digital domain, and the converter can start its normal ADC operation.

In order to implement the measurement technique on-chip, it has to fulfill several constraints: it has to be reliable, accurate, small, and realizable on-chip. Moreover, it is undesirable to modify the DAC-core to support the measurement procedure by means of additional switches or sources, as this could influence (dynamic) performance of the DAC adversely. To comply with all these constraints, the setup of Fig. 4.5 is proposed. It uses a simple analog measurement circuit (composed of a band-pass filter (BPF) and a comparator), and a digital measurement algorithm that provides the digital input code to the DAC-core during the self-measurement. An important advantage of this setup is that it can measure the individual current sources by looking only at the overall (combined) output of the DAC. In that way, the method prevents the need for access to the individual current sources, which would complicate the circuit design.

Measurement algorithm

The measurement algorithm aims at minimizing analog circuitry of the measurement technique by using digital algorithms as much as possible. Instead of measuring the values of the current sources in an absolute sense (which would require an accurate ADC), sources are measured relatively to each other only.

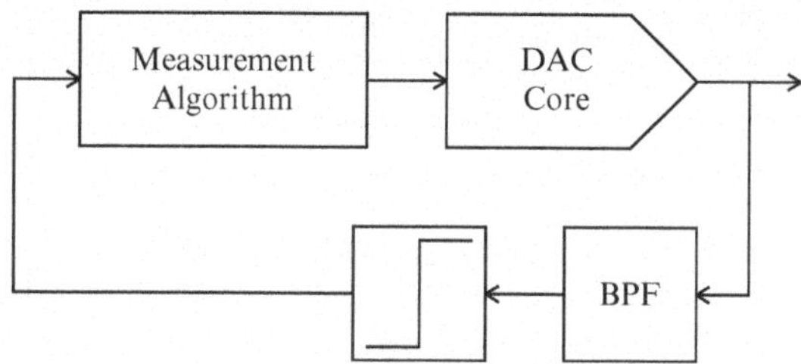

Figure 4.5. Detailed view of the built-in measurement setup, composed of a band-pass filter, a comparator and an algorithm.

The idea of the method is to find for each current source j a combination of current sources 0 up to $j-1$, of which the combined output current $I_{sum,j}$ approximates the actual current I_j of source j as good as possible. As the measurement is a *relative* measurement, the actual values of I_j and $I_{sum,j}$ are not important, it is sufficient to find a combination of sources which properly approximates I_j: $|\Delta_j| = |I_j - I_{sum,j}| \approx 0$. $I_{sum,j}$ can be written as:

$$I_{sum,j} = \sum_{i=0}^{j-1} S_{i,j} \cdot I_i \,, \tag{4.17}$$

where I_i is the actual current of source i, and $S_{i,j} = 0$ when source i is not used and $S_{i,j} = 1$ when source i is used in the combination approximating I_j. The combination of sources composing $I_{sum,j}$ can be found using a comparator determining the sign of Δ_j, and a successive-approximation algorithm minimizing $|\Delta_j|$ by controlling the current sources. In the actual design, a BPF was added to the analog circuit as will be explained in the next section. The measurement algorithm determines the values of $S_{i,j}$, based on which the digital representation ω_j of each current source j can be derived:

$$\omega_j = \sum_{i=0}^{j-1} S_{i,j} \cdot \omega_i \tag{4.18}$$

The measurement algorithm starts with initializing the digital representation of the smallest source (source 0) ω_0 to 1, an arbitrary unit value. Then, iteratively for all other sources j, starting with source 1, up to source $k-1$, the measurement procedure determining $I_{sum,j}$ is performed, and the digitized estimation ω_j can be derived. The algorithm determining ω_j is illustrated in Fig. 4.6. This algorithm is performed iteratively for the sources 1 up to $k-1$.

At the end of the measurement loop, all weights are scaled to normalize the range to the full-scale range of the N-bit input-code.

```
Turn off all sources (I_sum = 0)
for i = j − 1 down to 0
        Turn on source i (I_sum = I_sum + I_i)
        if I_sum > I_j
                S_i = 0
                Turn off source i (I_sum = I_sum − I_i)
        else
                S_i = 1
        end if
end for
```

Figure 4.6. Algorithm, finding a combination of sources approximating source j.

Analog measurement circuit

The analog part of the measurement setup has to provide the digital algorithm with the sign information of Δ_j. In [31], a two-step approach is used. First the value of I_j is recorded on a variable current source (implemented as a sub-binary DAC). In the second step, the recorded value is compared to I_{sum}, yielding the sign of Δ_j. The main disadvantage of this method is that it requires a complete DAC, of which the accuracy limits the accuracy of the measurement. Moreover, due to this implementation, the unit elements in the DAC-core have to be disconnected from the normal output and reconnected to the measurement circuitry by means of a switch, which could affect the performance. Therefore, another approach is proposed here requiring neither an additional DAC in the measurement setup nor a switch inside the DAC-core as it measures the voltage rather than sensing the current.

It is assumed that the current-steering DAC is designed as a differential DAC, which is the case for almost any high-performance DAC nowadays. In this situation, the DAC has a positive and a negative output, and the current of each source j is connected to either of the two outputs. During the measurement of source j, it is not possible to disconnect the sources not taking part in the algorithm (i.e. all sources i with either $S_{i,j} = 0$ or $i > j$). In order to distinguish the information to be measured from the superfluous information, the information to be measured is modulated on a carrier signal: the DAC continuously alternates between two states ϕ_1 and ϕ_2. All sources not taking part in the measurement remain connected to the negative output of the DAC, resulting in a DC output current I_{DC}. Source j (the source of which the value has to be determined) is connected to the positive output during phase ϕ_1 and to the negative output during phase ϕ_2. All sources i with $S_{i,j} = 1$ are connected to the negative output during phase ϕ_1 and to the positive output during phase ϕ_2. Figure 4.7 illustrates the output of the DAC as a function of time.

When a band-pass filter is connected to the output of the DAC, the DC level is blocked, and hence the comparator is provided only by the information to

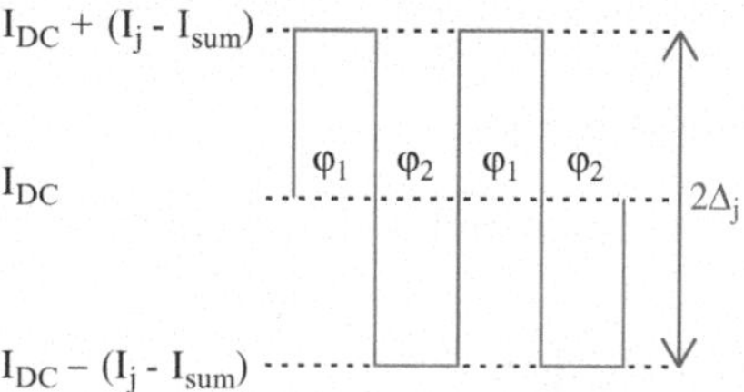

Figure 4.7. DAC output as a function of time during the measurement of source j.

be measured (the modulated '$I_j - I_{sum}$'-signal). The output of the comparator becomes a square wave, of which the phase (with respect to ϕ_1 and ϕ_2) corresponds to the sign of Δ_j, which is the information required by the digital measurement algorithm. The band-pass filter also rejects high-frequency components, to make the system less prone to noise and interference.

4.4 Digital pre-correction algorithm

The digital pre-correction algorithm has to select a suitable combination of current sources for each possible input code, based on the measurement results of the actual values of the sources (the values ω_j). A successive-approximation algorithm is used to find a suitable combination: starting with the largest source $j = k - 1$, the values ω_j are either added to or subtracted from the input code, such that the residual value is minimized. Corresponding to this addition or subtraction, the actual source is connected to the negative or the positive output of the DAC respectively. Figure 4.8 illustrates the algorithm.

$residue \leftarrow$ 'new input code'
for $i = k - 1$ down to 0
    if $residue \geq 0$
        select w_i positive
        $residue = residue - w_i$
    else
        select w_i negative
        $residue = residue + w_i$
    end if
end for

Figure 4.8. Digital pre-correction algorithm.

5. Design example

In this section, a design example for a 12-bit CSA is discussed. Three alternative solutions will be shown, namely:

- A binary-scaled design, based on intrinsic accuracy.

- A sub-binary variable-radix design.
- A sub-binary fixed-radix design.

The three solutions will be compared with respect to area and accuracy.

5.1 Binary-scaled design

From (4.2), it follows that for 12-bit accuracy a matching of $\sigma_u/I_u = 0.37\%$ is required. The current elements are simply binary-scaled, and the area of the unit element can be calculated according to (4.4). Table 4.4 shows the parameters of this DAC design, resulting in an overall CSA area of 0.1 mm^2.

Table 4.4. Binary-scaled intrinsic design.

Source i	Relative size α_i	Radix ρ_i	Area WL (μm^2)
0	1	–	24
1	2	2	48
2	4	2	96
3	8	2	192
4	16	2	384
5	32	2	768
6	64	2	1.5 k
7	128	2	3.1 k
8	256	2	6.1 k
9	512	2	12 k
10	1,024	2	25 k
11	2,048	2	49 k
Σ	**4,095**	–	**98 k**

5.2 Sub-binary variable-radix design

For a sub-binary variable-radix design, the equations from (4.15) determine the relative value of the current elements. As opposed to the intrinsic design, where σ_u/I_u is fixed by the required accuracy, in this case, σ_u/I_u is a design parameter that can be chosen freely. As a rather extreme example, $\sigma_u/I_u = 7.5\%$ is selected here. This poor matching corresponds to a transistor size of only $0.06\ \mu\text{m}^2$, which is almost the minimum possible size in a $0.18\ \mu\text{m}$ technology. Given this standard deviation and a confidence-level $\lambda = 4$, the current sources can be calculated from (4.15), while the area is given by (4.4). The results, also showing the variable-radix, are given in Table 4.5. A total of 16 sources is required to achieve 12-bit performance because of the sub-binary design, as can be verified from the simulation results later in this chapter.

Table 4.5. Sub-binary variable-radix design.

Source i	Relative size α_i	Radix ρ_i	Area WL (μm^2)
0	0.7416	–	0.04
1	1.3117	1.77	0.08
2	2.4189	1.84	0.14
3	4.5702	1.89	0.26
4	8.7762	1.92	0.51
5	17.0473	1.94	0.98
6	33.3876	1.96	1.9
7	65.7806	1.97	3.8
8	130.1583	1.98	7.5
9	258.3352	1.98	15
10	513.8701	1.99	30
11	1023.7815	1.99	59
12	2041.9653	1.99	118
13	4076.0151	2.00	235
14	8140.8365	2.00	469
15	16265.8430	2.00	937
Σ	**32,585**	–	**1.9 k**

In this case, the total CSA area[2] equals 0.0019 mm^2, which is 50 times smaller than an intrinsic radix-2 design.

For illustration of the self-measurement and pre-correction algorithm, this design is now taken as an example. Suppose that the actual values of the sources of this converter correspond to the values as given in Table 4.5.

First, to initialize the recursive measurement algorithm, ω_0, the digital value representing source 0, is set to 1. Now, source 1 with value $\alpha_1 = 1.3117$ can be measured using the algorithm from Fig. 4.6 with $j = 1$. The only source that can be used to compose I_{sum} is source $i = 0$. Turning this source on yields $I_{sum} = \alpha_0 = 0.7416$. As $I_{sum} \leq I_j$ the algorithm results in $S_0 = 1$, and thus the digital representation of source 1 becomes $\omega_1 = \omega_0 = 1$ (according to (4.18)).

In the next step, the algorithm is repeated for source $j = 2$ with value $\alpha_2 = 2.4189$. First, I_{sum} is set to $\alpha_1 = 1.3117$. As I_{sum} is smaller than I_j, S_1 becomes 1. Then, α_0 is added to I_{sum}. As I_{sum} is still smaller than I_j, S_0 becomes also 1, and $\omega_2 = \omega_1 + \omega_0 = 2$ is yielded.

Likewise, the algorithm is repeated for the other sources composing the converter, resulting in the measured values ω_i as given in Table 4.6. Also, as the 12-bit digital input code has a range of [−2048:2047], the weights need to be scaled such that their range corresponds to the input-code range. These normalized values are also shown in the table.

Table 4.6. Measured values of the current sources of the sub-binary variable-radix design.

Source i	**Real value** α_i	**Measured value** ω_i	**Normalized measured value** $\omega_{norm,i}$
0	0.7416	1	0.0778
1	1.3117	1	0.0778
2	2.4189	2	0.1556
3	4.5702	4	0.3112
4	8.7762	7	0.5446
5	17.0473	14	1.0891
6	33.3876	27	2.1004
7	65.7806	53	4.1231
8	130.1583	105	8.1684
9	258.3352	209	16.2589
10	513.8701	415	32.2844
11	1023.7815	827	64.3355
12	2041.9653	1650	128.3598
13	4076.0151	3293	256.1750
14	8140.8365	6577	511.6499
15	16265.8430	13141	1022.2885

Next, the operation of the digital pre-correction algorithm is illustrated, using the obtained normalized weights from Table 4.6. Suppose an input-code equal to 827 is applied and needs to be converted. Following the algorithm from Fig. 4.8, the $residue$ is set to 827. As the $residue$ is larger than zero, the first weight ($\omega_{norm,15}$) is selected positive and the new residue becomes $residue = 827 - \omega_{norm,15} = -195.3$. Now, $residue$ is negative, thus in the next iteration $\omega_{norm,14}$ will be selected negative and the new residue becomes $residue = -195.3 + \omega_{norm,14} = 316.4$. This process is repeated for all 16 sources. Table 4.7 summarizes the iterations of the algorithm.

Finally, the actually generated output is given by the obtained settings of the switches S_i, and the real values of the current sources α_i:

$$I_{out} = \sum_{i=0}^{k-1} S_i \cdot \alpha_i \,, \tag{4.19}$$

which equals 13,157 (unit element currents) in this case. The full-scale current is given by the summation over all α_i, which is 32,585 (unit element currents). The actual output as a ratio of the full-scale is thus 13,157/32,585 = 40.38%. This corresponds to the required output as the digital input-code 827 is also 40.38% of the full-scale (2,048).

Table 4.7. Example of digital pre-correction for input code 827.

Iteration	*Residue*	*Residue* ≥ 0 ?	Weight contribution	
1	$+827.0000$	y → $S_{15} = +1$	$\omega_{norm,15}$	1022.2885
2	-195.2885	n → $S_{14} = -1$	$\omega_{norm,14}$	−511.6499
3	$+316.3614$	y → $S_{13} = +1$	$\omega_{norm,13}$	256.1750
4	$+60.1864$	y → $S_{12} = +1$	$\omega_{norm,12}$	128.3598
5	-68.1734	n → $S_{11} = -1$	$\omega_{norm,11}$	−64.3355
6	-3.8380	n → $S_{10} = -1$	$\omega_{norm,10}$	−32.2844
7	$+28.4465$	y → $S_9 = +1$	$\omega_{norm,9}$	16.2589
8	$+12.1876$	y → $S_8 = +1$	$\omega_{norm,8}$	8.1684
9	$+4.0192$	y → $S_7 = +1$	$\omega_{norm,7}$	4.1231
10	-0.1039	n → $S_6 = -1$	$\omega_{norm,6}$	−2.1004
11	$+1.9966$	y → $S_5 = +1$	$\omega_{norm,5}$	1.0891
12	$+0.9075$	y → $S_4 = +1$	$\omega_{norm,4}$	0.5446
13	$+0.3629$	y → $S_3 = +1$	$\omega_{norm,3}$	0.3112
14	$+0.0517$	y → $S_2 = +1$	$\omega_{norm,2}$	0.1556
15	-0.1039	n → $S_1 = -1$	$\omega_{norm,1}$	−0.0778
16	-0.0261	n → $S_0 = -1$	$\omega_{norm,0}$	−0.0778
	$+0.0517$		$\Sigma\omega$	826.9483

5.3 Sub-binary fixed-radix design

For the design with redundancy and a fixed radix, σ_u/I_u is set to 7.5% as before. In order to meet the confidence-level of $\lambda = 4$ for all cases, the radix was set to $\rho = 1.77$, which corresponds to the ρ_1 of the variable-radix design. As a result, α_0 and α_1 also remain equal to the previous case. However, for the other sources, the values start to deviate as the radix remains constant in this case. As a result of the reduced radix, more current sources are required for this design to achieve sufficient dynamic range, leading to an overall area of 0.005 mm^2, being 2.5 times as much as for the variable-radix design. Table 4.8 lists the values of the current sources for the fixed-radix design.

5.4 Comparison of performance

To verify that the three alternative designs all achieve their 12-bit accuracy target, Monte-Carlo simulations were carried out on 10,000 samples of each converter. Mismatch was added to each unit-source, corresponding to $\frac{\sigma_u}{I_u} = 0.37\%$ for the intrinsic design and $\frac{\sigma_u}{I_u} = 7.5\%$ for the other two designs. For the two sub-binary designs, an implemented measurement algorithm was used to determine the digital coefficients representing the values of the current sources, based on which the digital pre-correction algorithm was applied. Figure 4.9 shows the achieved INL and DNL for all converters. Most converters achieve the 0.5 LSB target for INL and DNL. The variable-radix design achieves the best performance, revealing that it is somewhat overdesigned compared to the

Table 4.8. Sub-binary fixed-radix design.

Source i	**Relative size** α_i	**Radix** ρ_i	**Area** WL (μm^2)
0	0.7416	–	0.04
1	1.3117	1.77	0.08
2	2.3201	1.77	0.13
3	4.1036	1.77	0.24
4	7.2580	1.77	0.42
5	12.8374	1.77	0.74
6	22.7057	1.77	1.3
7	40.1599	1.77	2.3
8	71.0313	1.77	4.1
9	125.6339	1.77	7.2
10	222.2100	1.77	13
11	393.0253	1.77	23
12	695.1482	1.77	40
13	1229.5162	1.77	71
14	2174.6586	1.77	125
15	3846.3424	1.77	222
16	6803.0675	1.77	392
17	12032.6593	1.77	693
18	21282.2950	1.77	1.2k
19	37642.2261	1.77	2.2k
Σ	**86,609**	–	**5.0 k**

other designs, i.e. the area of the variable-radix could be further reduced while maintaining 12-bit accuracy.

An overview of the specifications, including the CSA area and the required number of independent current sources is given in Table 4.9.

Table 4.9. Comparison of DAC performances.

	Accuracy (bit)	**Transistor area (mm^2)**	**Number of sources**
Binary intrinsic design	12	0.1	12
Sub-binary variable-radix design	12	0.0019	16
Sub-binary fixed-radix design	12	0.005	20

From the table, the following conclusions can be drawn for this 12-bit example:

- The variable-radix design achieves the smallest area, about 50x smaller compared to the intrinsic design and also substantially smaller than the fixed-radix design.
- The variable-radix design requires only 16 sources, which is 4 more compared to an intrinsic design and 4 less compared to the fixed-radix design.

None of the current-source calibration methods from Section 3 can achieve this small overhead in number of elements.

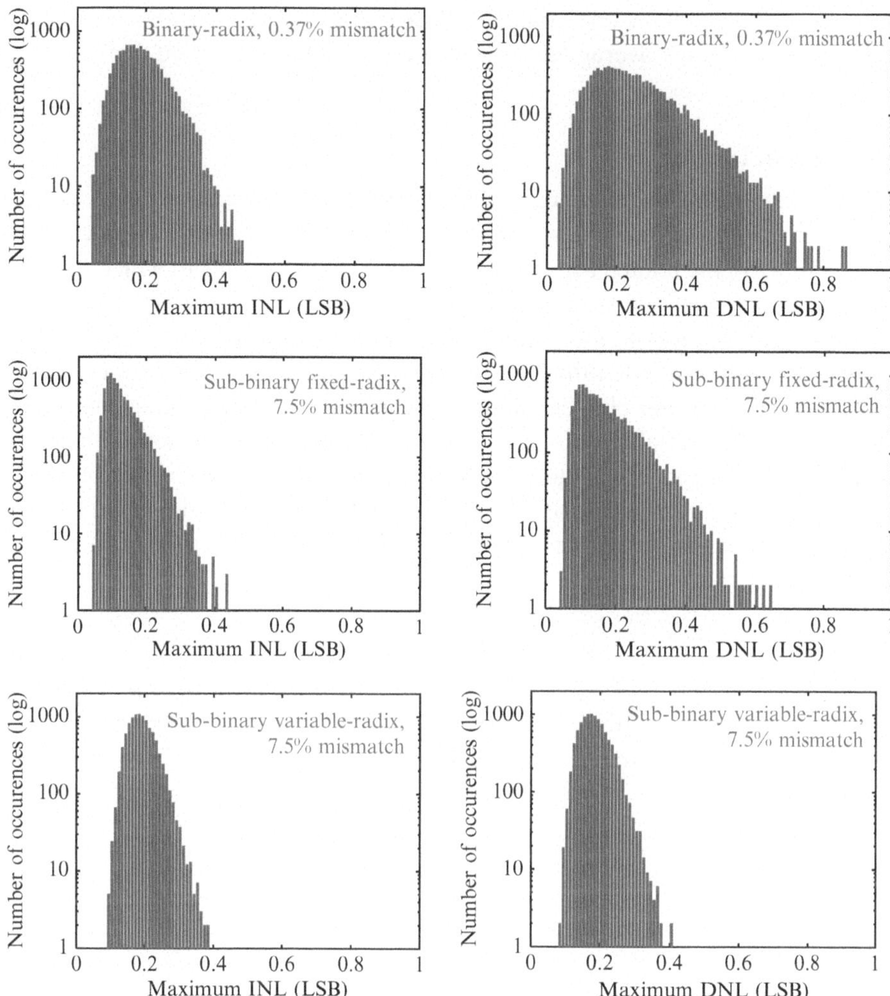

Figure 4.9. Histogram of maximum INL (left) and maximum DNL (right) for 10,000 samples of an intrinsic design (top), a sub-binary fixed-radix design (middle) and a sub-binary variable-radix design (bottom).

For higher resolutions, there is even more advantage for a sub-binary variable-radix design. For example, consider a 13-bit DAC design: in case of an intrinsic-design, the area will increase with a factor of four compared to the 12-bit design. For the sub-binary variable-radix design, as the same 7.5% unit mismatch can be tolerated, one only needs to add more sources to achieve the new full-scale range. As for the largest sources the radix approximates two, one additional source will be sufficient. Thus, the area increase is a factor of two in this situation. The same reasoning holds for the sub-binary fixed-radix design. However, as the radix remains smaller than two in that case, one might need to add more than one source to achieve the required full-scale range, thus achieving on the average an area increase of somewhat more than a factor of two.

6. Conclusion

In this chapter, the smart concept was applied to DA converters with as aim to minimize the analog area as much as possible, and to use digital processing instead to solve the related accuracy problem. A few examples in which this approach can be beneficial were highlighted. From a literature study and a mathematical analysis, it was shown that area is an issue for high-resolution DAC design. In order to reduce the overall area both the area of the unit elements and the number of unit elements have to be minimized. A new concept, based on a sub-binary variable-radix current-source-array was developed that can achieve both targets. The required self-measurement and self-correction technique were explained and verified with a design example. In the next chapter, the theory of a sub-binary variable-radix design will be verified with a chip implementation.

Notes

1. Note that in this work, *resolution* refers to the number of digital bits of the converter, without referring to the actual accuracy of the converter.
2. Considering active transistor-area only, excluding overhead from interconnect, spacing, etc.

Chapter 5

DESIGN OF A SUB-BINARY VARIABLE-RADIX DAC

This chapter discusses the chip-implementation and experimental verification of a sub-binary variable-radix DAC, according to the theory presented in Chapter 4. The circuit-level design is described in Section 1 and the layout in Section 2. The off-chip implementation of the self-measurement circuit is shown in Section 3. Then, experimental results and conclusions are given in Section 4 and Section 5, respectively.

1. Schematic design

In this section, the circuit-level design of the DAC in a 0.18 μm CMOS technology is reviewed. Corresponding to the architecture from Fig. 4.1, the schematic was implemented as shown in Fig. 5.1. For this design, the digital pre-processor and the measurement circuitry were implemented off-chip. The main components (serial-in/parallel-out register, switch driver, switches and CSA) will be discussed next. The serial-in/parallel-out register operates at 1.8 V to comply with the external interface, but the DAC itself is designed to operate at a supply of 1.4 V only. The selection of the low supply has two reasons: first, it will prove that the DAC design allows low-voltage operation while maintaining sufficient linearity, which is a necessity for future low-voltage technologies. Secondly, the DAC will be used as an on-chip test-signal generator for the T&H (see Chapter 7). With the selected supply, the DAC generates a differential voltage swing of $1V_{pp}$ around a common-mode level of 1.15 V, which is suitable for the T&H.

1.1 Serial-in parallel-out register

Because of pin limitations, the 16-bit data-input is provided through a serial interface: 16 clock pulses on the *Serial clock* input are required to load a 16-bit

P. Harpe et al., *Smart AD and DA Conversion*,
Analog Circuits and Signal Processing, DOI 10.1007/978-90-481-9042-3_5,

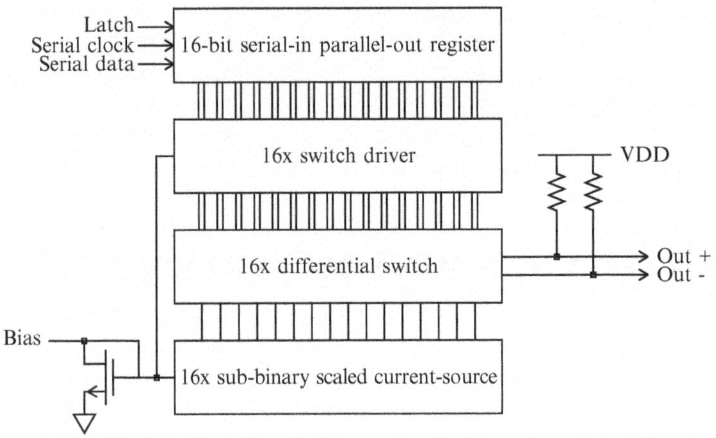

Figure 5.1. Schematic of the implemented current-steering DAC.

input code through *Serial data* into the 16-bit serial-in/parallel-out register. When the data is loaded into the register, it will be stored internally, but it is not yet provided to the switch driver. As soon as a positive clock-edge on *Latch* is provided, the internally stored vector will be made available to the switch driver. In that way, the *Latch* acts as a master-clock for the DAC. To be compliant with the external interface, the register operates at 1.8 V supply. Because of the serial interface, the sampling rate of the DAC is limited: the serial-in parallel-out register allows a maximum sampling rate of 11MHz, but in practice, the measurement setup limits the sampling rate to 1.5 MHz.

1.2 Switch driver

Sixteen switch drivers are needed to drive the 16 differential switches controlling the current sources. Because of the speed-limitations of the serial register, the DAC cannot be tested at high operating frequencies. Therefore, the switch drivers were not optimized for dynamic performance, but only for static performance. The schematic of a switch driver is shown in Fig. 5.2. By selecting the proper IR-drop, the output level of the switch driver is either 1.4 V to turn a switch on, or 0.4 V to turn a switch off. The required bias for the tail current source is derived from the internal current mirror, that also provides the biasing for the CSA (see Fig. 5.1).

1.3 Switches

The analog switches are used to connect each of the current sources from the CSA to the positive or the negative output of the DAC. The implementation is shown in Fig. 5.3: the gate-inputs are controlled by the differential switch-driver, while the tail is connected to one of the current sources in the CSA.

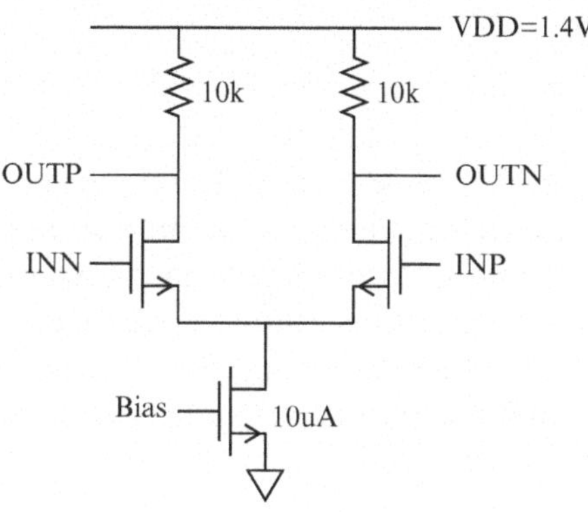

Figure 5.2. Schematic of the switch driver.

The drains of the switches are connected together to the output of the DAC to sum the output currents. Also (Fig. 5.1), on-chip termination resistors of 75 Ω are included to convert the output current into a voltage. The resistors are implemented on-chip, as one of the applications of the DAC is to use it as an on-chip test-signal generator. Then, it is convenient to prevent the need for external resistors. As the switches are switching an analog signal, for best linearity their W/L should scale proportional to the current being switched. With that goal, the switches are implemented with the same transistor sizes as the current sources themselves; i.e. the switches are also sub-binary scaled.

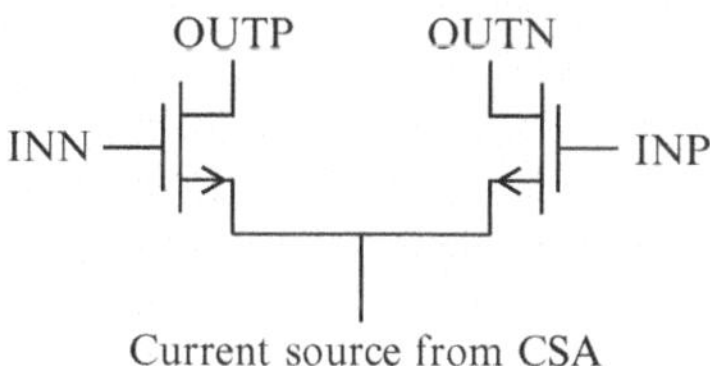

Figure 5.3. Schematic of the switch.

An important feature of this design to enable 1.4 V operation (for a 0.18μm technology) is to implement no cascode transistors at all: nor for the switches, nor for the current sources. Normally, it is assumed that a high output resistance of the current elements and switches is required to achieve sufficient overall linearity of the DA conversion [32]. However, the simulation results later in this chapter confirm that with small-length devices and without cascodes, even at 1.4 V supply and a $1V_{pp}$ output swing, the output resistance is sufficient for the overall linearity requirement.

1.4 CSA design

For the CSA design, the sub-binary variable-radix approach from Chapter 4 was used. Apart from the area requirement, there is also a minimum-length requirement for each current source to meet the output resistance specification. From simulations, it follows that a length of 2 μm is sufficient. Then, given the length and the area requirements according to the previously discussed design example, the width of the transistors can be calculated. Table 5.1 shows the previously calculated area requirement and the implemented W, L and area for each device. When multiple transistors are used in series or in parallel, this is indicated by a multiplication factor for the length or width, respectively.

For sources 0 up to 4, the calculated W becomes smaller than the smallest W that can be implemented in the 0.18 μm-technology. Because of that, for those sources the W is set to the minimum value of 0.26 μm and the length is increased to achieve the required W/L. Due to the increase of L, the area of these sources becomes larger than strictly required. Nonetheless, the effect can be neglected as the overall area of 0.0019 mm^2 is only 1.5% larger than the expected area.

Table 5.1. Theory and implementation of the CSA.

		Theory	Implementation		
Source	Relative size	Area	Width	Length	Area
i	α_i	WL_{min} (μm^2)	W (μm)	L (μm)	WL (μm^2)
0	0.7416	0.04	0.26	$7 \cdot 3.30$	6.08
1	1.3117	0.08	0.26	$4 \cdot 3.30$	3.43
2	2.4189	0.14	0.26	$2 \cdot 3.58$	1.86
3	4.5702	0.26	0.26	3.80	0.99
4	8.7762	0.51	0.26	1.98	0.51
5	17.0473	0.98	0.52	2.00	1.0
6	33.3876	1.9	1.00	2.00	2.0
7	65.7806	3.8	1.98	2.00	4.0
8	130.1583	7.5	3.90	2.00	7.8
9	258.3352	15	7.76	2.00	16
10	513.8701	30	$2 \cdot 7.70$	2.00	31
11	1023.7815	59	$3 \cdot 10.16$	2.00	61
12	2041.9653	118	$6 \cdot 10.06$	2.00	121
13	4076.0151	235	$12 \cdot 9.96$	2.00	239
14	8140.8365	469	$24 \cdot 9.86$	2.00	473
15	16265.8430	937	$46 \cdot 10.18$	2.00	937
Σ	**32,585**	**1.88 k**	–	–	**1.90 k**

A potential issue with the selected dimensions is that the transistors are not based on multiples of a unit-element. This could cause systematic mismatch errors on top of the random mismatch errors. Lacking information about the magnitude of these systematic mismatch errors, the following approach is used: first, abundant redundancy is added (7.5% with a 4σ confidence-level for the unit element). As the random errors are most likely smaller than that, there is margin left for systematic errors. Then, as some of the systematic errors are taken into account in the transistor models, Cadence simulations are used to verify the redundancy for the transistors. As a last step, the experimental results can give insight in this relatively unknown topic.

1.5 Simulation results

The complete DAC (Fig. 5.1) was simulated in Cadence to verify the performance in terms of INL/DNL. For convenience, the digital pre-processing was done off-line in Matlab. The power supply of the DAC is set to 1.4 V while the bias current is tuned to realize a $1V_{pp}$ output swing over the 75 Ω load resistors. The simulated INL/DNL curves are shown in Fig. 5.4. As INL and DNL remain within 0.5 LSB, the 12-bit accuracy is confirmed. Note that the INL is actually limited by the output resistance and not by the current-source matching. However, the simulations reveal that the output-resistance requirement can be achieved given the low supply, large swing and the lack of cascodes.

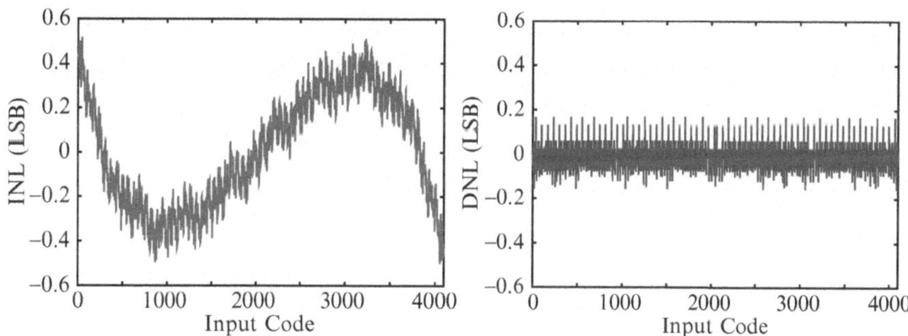

Figure 5.4. Cadence simulation of the sub-binary variable-radix DAC: INL (left) and DNL (right); $V_{DD} = 1.4$ V, $V_{pp,out} = 1.0$ V.

2. Layout

The layout of the sub-binary variable radix DAC is shown in Fig. 5.5. The 16 elements in the CSA are laid out next to each other, in a single row. For symmetry, the differential switches (of which the layout is identical to the layout of the CSA) surround the CSA.

The total area (excluding the self-measurement circuitry and pre-correction circuitry) is 0.03 mm^2, while the CSA occupies 0.005 mm^2. Table 5.2 gives an overview of the area of the most important sub-components. The CSA

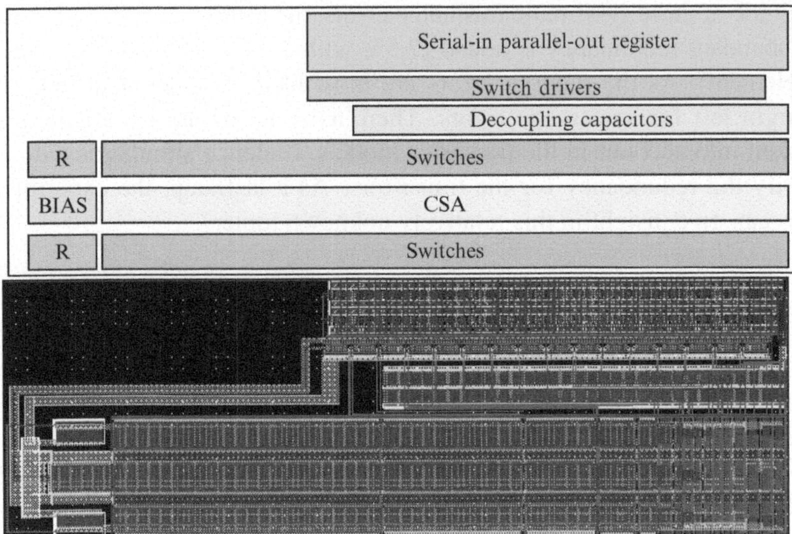

Figure 5.5. Floorplan and layout of the DAC. The shown area is 350 × 115 μm.

area[1] is at least 10× smaller compared to alternative corrected CSA structures (see Table 4.3), showing that the approach for minimizing the CSA area was successful.

Table 5.2. Area of the various components inside the DAC.

Component	Area (mm^2)
CSA	0.005
Switches	0.010
Switch drivers	0.002
Serial-in parallel-out register	0.005
Remaining components, wiring, spacing	0.008
Total DAC	**0.030**

3. Self-measurement-circuit implementation

For the experimental verification of the DAC, a self-measurement circuit is needed. As there is none implemented on-chip, an external circuit has been made with discrete components as shown in Fig. 5.6.

Corresponding to the idea from Section 4.3, the circuit implements a band-pass filter and a comparator. By adding gain to the two-stage filter, the offset of the comparator is effectively reduced. The offset reduction will simplify an on-chip implementation, as it relaxes the accuracy constraints. For this discrete

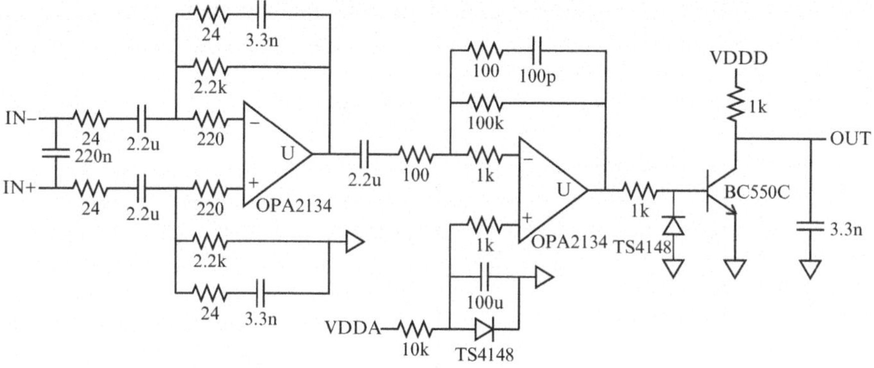

Figure 5.6. Schematic of the discrete self-measurement circuit.

prototype, the pass-band of the BPF was set to 5kHz, as it is simple to realize with off-the-shelf components. For an on-chip implementation, a much higher frequency would be advantageous as it reduces the values of the required capacitors and thereby the chip area.

4. Experimental results

4.1 Measurement setup

As the self-measurement and pre-correction are not implemented on-chip, the experimental measurements on the DAC are carried out in three steps:

1. Self measurement of the current sources.
2. Determination of the transfer curve.
3. Digital pre-correction of the transfer curve.

The first step is to characterize the current sources by means of the self-measurement method. For that goal, the setup depicted in Fig. 5.7 is used that implements the method described in Section 4.3. The detailed equipment setup is shown in Fig. 5.8. A Matlab program runs the recursive measurement algorithm autonomously. The sources to be measured are selected from the PC through the FPGA-interface. The off-chip measurement circuit determines the outcome of the comparison, resulting in a modulated binary output (see Fig. 4.7). The output is demodulated by the FPGA and provided to a multimeter that captures the outcome and sends it back to the PC to close the measurement loop. Note that the multi-meter only reads the digital output decision of the FPGA (0 or 1), it does not perform an analog measurement. The reason to include the multi-meter is because it can be controlled conveniently through a GPIB interface. At the end of the self-measurement procedure, all the weights of the current sources are known in the digital domain, and stored into Matlab.

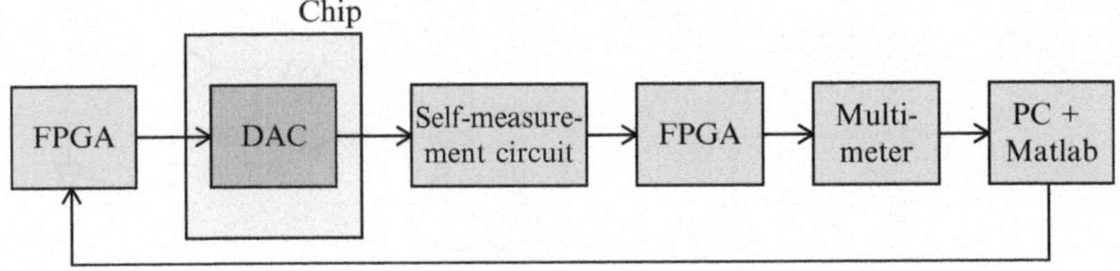

Figure 5.7. Self-measurement setup.

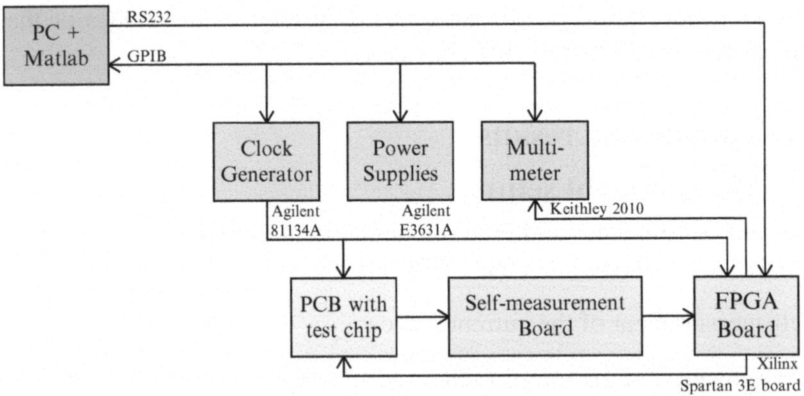

Figure 5.8. Equipment setup for the self-measurement phase.

The second step in the series of measurements is to acquire the transfer-curve of the DAC, which is done by the setup of Fig. 5.9; the equivalent equipment setup is given in Fig. 5.10.

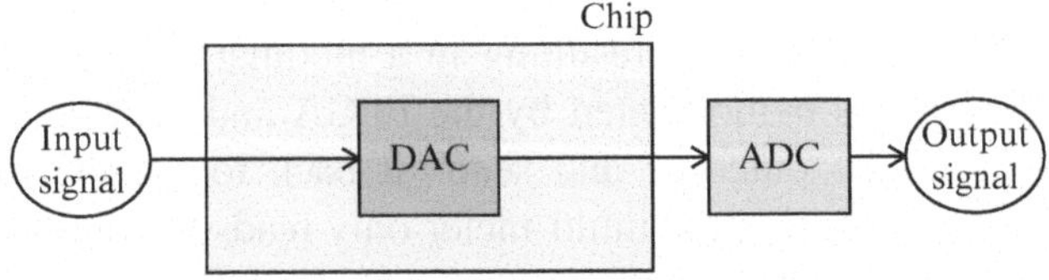

Figure 5.9. Setup for the determination of the transfer-curve.

As the pre-correction is performed off-line in Matlab, the complete uncorrected transfer curve is measured and stored: all 2^{16} combinations of

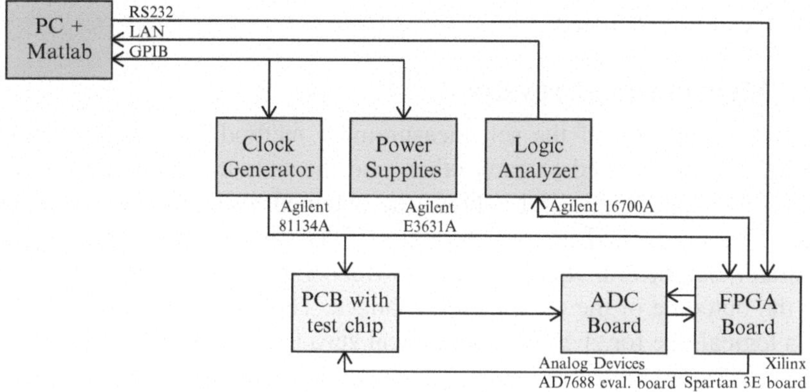

Figure 5.10. Equipment setup for measuring the transfer curve.

current sources are selected one by one, digitized by an off-chip ADC, and stored on the PC. At the end, this results in a complete lookup-table describing all possible output currents of the DAC. Figure 5.10 shows a few details of the setup: initiated by the PC, the FPGA will produce a 16-bit digital code-sweep for the DAC. Also controlled by the FPGA, the ADC digitizes the DAC output and the outcome is stored into the logic analyzer. Then, the data stored in the logic analyzer is captured by the PC for further processing in Matlab.

The third step of the experiments is to apply the digital pre-correction algorithm. In the previous two steps, the weights and the uncorrected transfer curve were measured and stored in Matlab. With this information, the pre-correction can be performed in Matlab as shown in Fig. 5.11. A 12-bit input

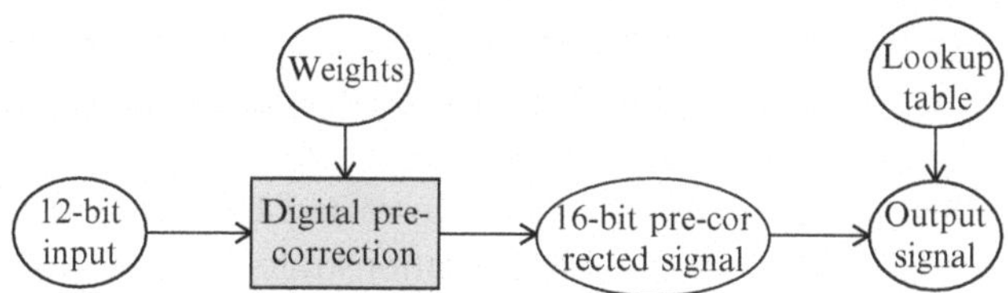

Figure 5.11. Digital pre-correction setup (software).

ramp is created, corresponding to the target resolution of the corrected DAC. With the previously measured weights and the pre-correction algorithm from Section 4.4, the 12-bit ramp is mapped to a 16-bit pre-corrected signal. As the actual transfer curve of the DAC is already stored in a lookup-table, the 16-bit pre-corrected signal simply indexes the lookup-table to find the corresponding output levels of the DAC. In the end, this part of the experiments yields the 12-bit digitally pre-corrected transfer curve of the DAC, from which the INL and DNL can be calculated.

4.2 Measurement results

As a first verification of the self-measurement method, two current sources (A and B) are compared to each other, and the output signals are observed on an oscilloscope. Figure 5.12 shows the obtained results for two situations: in case 1 $A > B$, and in case 2 $A < B$. As can be seen, the modulated comparator output is a square wave of which the phase changes with 180° when the outcome of the comparison changes. Thus, the demodulated output gives a logical one for $A > B$ and a logical zero for $A < B$.

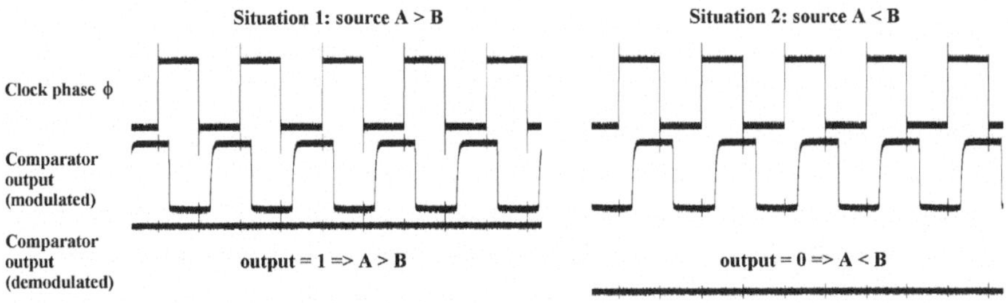

Figure 5.12. Example of the self-measurement of two sources.

As a next step, the autonomous self-measurement loop from Fig. 5.7 was initiated. Table 5.3 shows the decisions made by the comparator and the obtained weights for the first 10 sources of a test-chip. The measurement of source 9 indicates a problem in the chip: all comparator decisions are 1 in this case, which means that the summation of the sources 0 up to 8 is still less than the value of source 9. This implies that there is no redundancy at this transition and thus there will be a gap in the transfer curve between binary code 01.1111.1111 and 10.0000.0000. As the gap cannot be corrected by means of digital pre-correction, this test-chip is not suitable to demonstrate the method. Three other chips were also tested, all showing the same issue with current source 9. As the redundancy was designed for a 4σ confidence level, and the same problem occurs for four devices, there has to be a dominant systematic mismatch error that causes the gap for source 9. As explained in Section 1.4, it is obvious that systematic errors can be expected: the transistors are not based on unit cells. Thus, because of a.o. corner, edge and well-proximity effects, the intended ratio between two devices can be substantially different from the effective ratio. Unfortunately, these effects are not modeled for the given technology and can

Table 5.3. Self-measurement of a single DAC.

Source	**Decisions**									**Weight**
i	S_8	S_7	S_6	S_5	S_4	S_3	S_2	S_1	S_0	ω_i
0										1
1									1	1
2								1	1	2
3							1	1	1	4
4						1	0	1	1	6
5					1	0	1	1	1	10
6				1	1	0	0	1	1	18
7			1	1	0	1	1	0	1	35
8		1	1	0	1	1	0	0	1	74
9	1	1	1	1	1	1	1	1	1	151

be known only after characterization of the devices. On the other hand, with the information that is now available, this issue can be prevented by one of the following solutions:

- As a first solution, the unit-cell approach could be included in the design of the current sources to prevent systematic errors. That means that the α_i-ratios should become integer multiples of a unit-size. As for the smallest current sources the random mismatch will be dominant over the systematic mismatch, the unit-element approach is only necessary for the larger sources. The unit-element approach can be included in the design phase of the α_i-ratios by truncating each calculated α_i-value to the nearest multiple of unit-elements. Then, sufficient redundancy for each step is still guaranteed while systematic errors are prevented by the unit-element approach.

- As a second solution, the systematic mismatch errors can be modeled and taken into account in the design procedure. At the moment, the design procedure (equation (4.11)) takes only random mismatch into account. A second mismatch term for systematic errors could be added to these equations. Then, the obtained α_i-values will exhibit more redundancy to cover both types of mismatch.

To verify the digital pre-correction method with the existing chips, a work-around using two DACs is used here. Figure 5.13 shows how two DACs from two different chips are connected to effectively compose a single DAC: the outputs of the DACs are tied together to sum their individual output currents. The 16-bit input code d<15:0> is split in two parts: bits d<15:9> are applied to the corresponding bits of the first DAC, and bits d<8:0> are applied to the corresponding bits of the second DAC. The remaining inputs (bit <8:0> of DAC 1 and bit <15:9> of DAC 2) are tied to a constant logical level. Table 5.4 lists an overview of the connections, showing for each source α_i of each DAC whether it is connected to the digital input (d_i) or a constant logical level (0 or 1).

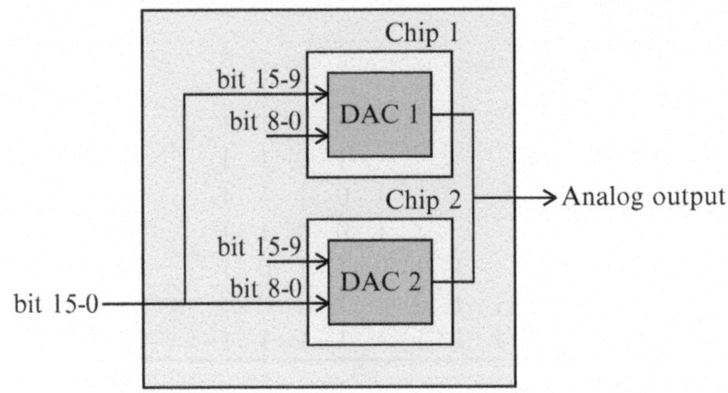

Figure 5.13. A DAC composed of two sub-DACs.

Table 5.4. Input mapping for the DAC composed of two sub-DACs.

	Source															
	α_{15}	α_{14}	α_{13}	α_{12}	α_{11}	α_{10}	α_9	α_8	α_7	α_6	α_5	α_4	α_3	α_2	α_1	α_0
DAC 1	d_{15}	d_{14}	d_{13}	d_{12}	d_{11}	d_{10}	d_9	1	0	0	0	0	0	0	0	0
DAC 2	1	0	0	0	0	0	0	d_8	d_7	d_6	d_5	d_4	d_3	d_2	d_1	d_0

As the two DACs are identical (except for random mismatch), the output current generated by this system is equivalent to the output current generated by a single DAC. However, as there are now two sets of load resistors connected in parallel, the output voltage swing is reduced by a factor of two in this dual-DAC design. The major advantage of the combination of two DACs is that each CSA has a separate bias current. Thus, the current sources of DAC 1 can be scaled relative to the current sources of DAC 2, i.e., the ratio of α_{15}–α_9 versus α_8–α_0 can be manually controlled by tuning the bias currents of the two DACs. From the previous self-measurement, it appeared that α_9 is too large compared to α_8–α_0. Now, by slightly increasing the bias current of α_8–α_0, the redundancy can be manually restored such that the pre-correction algorithm can be verified. Despite the unpleasant setup, this verification is still valuable as it will prove that 12-bit accuracy can be achieved using 16 inaccurately matched current sources. The fact that out of these 16 sources, 9 are located in one chip and 7 in another chip does not impede with that conclusion.

After connecting the two DACs together and tuning the bias to achieve redundancy for all sources, the self-measurement algorithm was executed again. The results of the self-measurement are shown in Table 5.5. For all sources (except source 1 and 2), there is at least one comparison that results in a zero.

Thus, for all sources there is redundancy available. For sources 1 and 2 there is also redundancy, but the redundancy is smaller than 1 LSB. As a consequence, the converter is still monotonous but the steps are smaller than 1 LSB. Because of the monotonicity the redundancy cannot be verified by the outcome of the self-measurement algorithm for these sources. From source 3 onwards, the redundancy results in non-monotonicity, which is reflected by the zero-decisions in the comparisons.

Table 5.5. Self-measurement of a DAC, composed of two sub-DACs.

Source								**Decisions**								**Weight**
i	S_{14}	S_{13}	S_{12}	S_{11}	S_{10}	S_9	S_8	S_7	S_6	S_5	S_4	S_3	S_2	S_1	S_0	ω_i
0																1
1															1	1
2														1	1	2
3													1	1	0	3
4												1	1	0	0	5
5											1	1	0	0	0	8
6										1	1	0	1	0	1	16
7									1	1	1	0	1	0	0	31
8								1	1	1	1	1	0	0	1	64
9							1	0	0	0	1	1	0	0	0	72
10						1	1	0	1	0	0	0	0	0	0	152
11					1	1	1	0	1	0	0	1	0	1	1	309
12				1	1	1	1	1	0	0	0	1	0	0	0	631
13			1	1	1	1	1	1	0	1	0	0	0	0	1	1,268
14		1	1	1	1	1	1	1	0	0	0	1	1	1	0	2,533
15	1	1	1	1	1	1	1	0	0	1	0	1	0	0	0	5,040

Having obtained the values of the weights, the next step is to determine the uncorrected transfer curve using the setup described in Fig. 5.9. Then, after storing the data to a lookup-table, the pre-correction is applied in Matlab as explained in Fig. 5.11 to obtain the corrected 12-bit transfer curve. For each point of the curve, four measurements are taken and averaged to reduce noise. From the corrected, averaged 12-bit transfer curve, the INL and DNL were derived, giving the results in Fig. 5.14. Both INL_{max} and DNL_{max} are around 0.4 LSB, yielding a 12.3 bit post-correction accuracy. Compared to the Cadence simulation (Fig. 5.4), there are two main differences: first, the global non-linearity in the INL curve is reduced by a factor of two. This is caused by the fact that for the dual-DAC measurement, the load resistance is reduced by a factor of two, thereby reducing the effect of the output resistance modulation of the current sources. A second difference is that in the measurement, the DNL performance is worse compared to the Cadence simulation. Several possible explanations are:

- The Cadence simulation is free of noise and disturbances. In reality, noise and disturbances are present and they can deteriorate the performance.

- In Cadence, the sources are ideal and mismatch-free. In reality, there is mismatch of the elements which can affect the performance.

- The specific systematic error that caused redundancy problems is not taken into account by the Cadence simulations. Though the redundancy could be restored by tuning the bias currents, the actual α_i's might be less optimal than the originally intended values, causing a loss of performance.

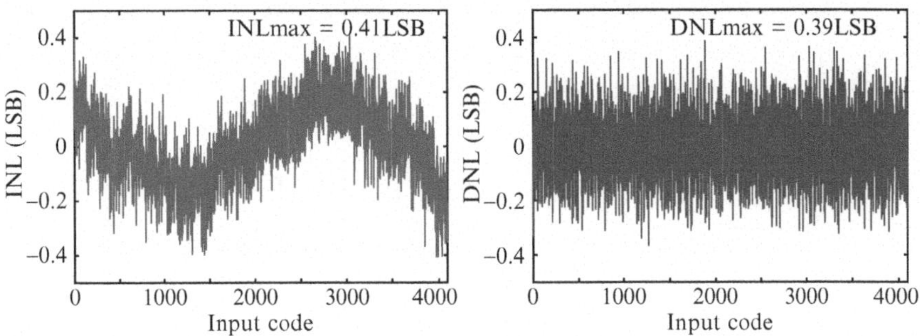

Figure 5.14. Measurement of the sub-binary variable-radix DAC: INL (left) and DNL (right); $V_{DD} = 1.4$ V. All points are measured four times and averaged.

5. Conclusion

In this chapter, a sub-binary variable-radix DAC was designed and implemented in a CMOS 0.18 μm technology. Because of the redundancy, the total CSA area (including layout-overhead) could be minimized to 0.005 mm^2, which is a 10$\times$ improvement over prior-art 12-bit converters (including overhead) and a 20$\times$ improvement compared to the intrinsic CSA area limitation (excluding overhead). To enable low-voltage operation, cascode devices were omitted completely. Nonetheless, sufficient output resistance could be achieved to reach over 12-bit linearity, which shows that most current designs are largely over-designed with respect to output resistance.

Because of systematic mismatch, the implemented DAC did not have sufficient redundancy to enable digital pre-correction of the imperfections. For future implementations, it is recommended to take the effect of systematic mismatch into account when designing the current elements. By doing so, sufficient redundancy for both random mismatch and systematic mismatch can be provided to ensure that the digital pre-correction can compensate for all the errors.

As a work-around, two DACs were connected together to validate the principle of operation and the achievable performance. The functionality of the self-measurement algorithm and pre-correction method was shown, and a final accuracy of 12.3 bit was measured.

Notes

1. Including overhead from e.g. interconnect, spacing, contacts, etc.

Chapter 6

SMART AD CONVERSION

This chapter applies the previously presented smart concept to analog-to-digital converters. A relevant overview of current state-of-the-art smart ADCs is given, and an approach is proposed how to apply the smart concept beneficially in this case. The actual proof-of-concept will be given in Chapters 7, 8 and 9, where the chip implementation is discussed.

1. Introduction

In Chapter 3, many different aspects and possible benefits of the smart concept were introduced. Concepts like e.g. self-test and correction were introduced and possible benefits with respect to e.g. performance, yield and portability were mentioned. In this chapter, a limited range within the smart concept will be selected for further investigation and actual chip implementation.

First of all, this project is focussing at high-speed, high-resolution AD converters, with pipelined converters in particular. Corresponding to the smart concept, it is assumed that the ADC is part of a mixed analog/digital system, such that, whenever necessary, the digital part can be used for processing of information. Nonetheless, it is recognized that the digital part does not come for free and has to be taken into account when evaluating the overall performance.

In this work, the smart concept will be applied with as main goal to improve the performance of a high-speed, high-resolution ADC. The performance is evaluated by means of a widely accepted figure-of-merit (FoM) that includes speed, accuracy and power consumption. The reason to focus on the FoM is because it is an important property of each ADC, independent on the application or situation where the ADC is used. Also, when the smart concept is able to improve the FoM, it inherently proves that the smart concept can be beneficial in a wide range of situations. As a second goal, this work aims at using circuit solutions that are portable to future technologies to provide a future-proof solution.

P. Harpe et al., *Smart AD and DA Conversion*,
Analog Circuits and Signal Processing, DOI 10.1007/978-90-481-9042-3_6,

This chapter starts with a literature study on current state-of-the-art in smart AD converter design in Section 2. Then, the approach to improve the performance using smart techniques will be introduced in Sections 3 and 4. Finally, the proposed approach is summarized in Section 5.

2. Literature review

Many publications exist that, in a certain way, include some of the aspects of the smart concept. However, by far most of these publications remain on the theoretical level, without an experimental verification. Without being complete, Tables 6.1 and 6.2 show an overview of recent publications with experimental results that use some form of smartness to improve the performance of high-speed ADCs. Table 6.1 shows in which way the smart concept is applied to the converter while Table 6.2 shows the achieved performance.

Most of the selected publications consider pipelined converters and correct for some of the errors inside the stages of the pipeline: offset, gain-error, distortion, incomplete settling of the amplifier or capacitor mismatch. Some of the publications consider a time-interleaved ADC and correct for mismatches between the channels of the converter. Most publications apply digital post-correction to correct for the imperfections, [45] and [48] also use analog correction. [37] and [43] use redundancy (in the capacitors or the channels) to improve the performance. On top of gain/offset calibration, [34] also includes adaptive biasing to optimize the power consumption for different situations.

The achieved performances are shown in Table 6.2. It should be noted that some publications provide measurement results for a relatively low input frequency (far below Nyquist) only. In these cases, the $ERBW$ is set to the maximum published frequency, as the performance for higher signal frequencies is unknown. The FoM is calculated according to:

$$FoM = \frac{Power}{2^{ENOB} \cdot min(f_s, 2ERBW)}, \text{ with } ENOB = \frac{SNDR - 1.76}{6.02} \tag{6.1}$$

The achieved FoM ranges from 0.12 pJ per conversion-step ([45]) up to 14.4 pJ per conversion-step ([34]). For comparison, Table 6.3 shows the performance of several state-of-the-art pipelined ADC designs that do not employ any calibration. These intrinsic designs achieve a FoM between 0.06 up to 0.8 pJ per conversion-step.

Figure 6.1 shows the performance of the calibrated designs from Table 6.2 and the intrinsic designs from Table 6.3. From this picture, it becomes clear that smart solutions are not necessarily better in FoM-performance than intrinsic designs. Though the smart designs prove their concept, namely the correction of specific imperfections, they do not prove that their approach enables an improvement of the overall performance.

Table 6.1. Overview of recent work on smart AD converters.

Reference	ADC type	Calibrated errors	Calibration method
[33], 2006	Pipelined ADC	Incomplete settling of stage	Digital
[34], 2006	Pipelined ADC	Gain/offset of stage, adaptive biasing	Digital, analog control
[35], 2006	Pipelined ADC	Gain and distortion of stage	Digital
[36], 2006	Pipelined ADC, time-interleaved	Stage mismatch of pipeline, no inter-channel correction	Digital
[37], 2007	Pipelined ADC	Capacitor mismatch	Capacitor redundancy
[38], 2007	Pipelined ADC	Incomplete settling, gain and distortion of stage	Digital
[39], 2007	Pipelined ADC, time-interleaved	Stage mismatch of pipeline, offset/gain between channels	Digital
[40], 2008	Pipelined ADC	Gain of stage	Digital
[41], 2008	Pipelined ADC, split ADC	Gain of stage, capacitor mismatch	Digital
[42], 2008	Pipelined ADC	Offset, gain, distortion of stage	Digital
[43], 2008	SAR ADC, time-interleaved	Offset/gain between channels	Channel redundancy
[44], 2008	Pipelined ADC	Gain and mismatch of stage	Digital
[45], 2009	Pipelined ADC	Offset and gain of stage	Digital and analog
[46], 2009	Pipelined ADC	Distortion of first stage	Digital
[47], 2009	Pipelined ADC	Gain of stage	Digital
[48], 2009	SAR ADC, time-interleaved	Gain, offset and timing between channels	Digital and analog
[49], 2009	Pipelined ADC	Capacitor mismatch	Digital
[50], 2009	SAR ADC, time-interleaved	Mismatch between channels	Digital
[51], 2009	Pipelined ADC	Capacitor mismatch and distortion	Digital
[52], 2009	Folding ADC	Offset	Digital
[53], 2009	Pipelined ADC	Gain, mismatch and distortion	Digital

In [34], [35], [38], [40] and [53], the power consumed by the digital post-calibration algorithm is either simulated or measured. With this information, a FoM that takes only the power of the digital calibration method into account can be calculated:

$$FoM_{calibration} = \frac{Power_{calibration}}{2^{ENOB} \cdot min(f_s, 2ERBW)}, \text{ with} \tag{6.2}$$

$$ENOB = \frac{SNDR - 1.76}{6.02},$$

where $Power_{calibration}$ is the power consumed by the digital post-calibration method. The results for the $FoM_{calibration}$ are given in Table 6.4, showing that the calibration costs about $50 - 200$ fJ/Conversion-step. From Table 6.3, it is known that a good design without calibration can achieve an overall

Table 6.2. Performance of recent work on smart AD converters.

Reference	Power (mW)	Sampling rate (MSps)	ERBW (MHz)	SNDR (dB)	FoM (pJ/conv.step)
[33], 2006	70	600	1100	25.5	7.6
[34], 2006	73	40	5	55.8	14.4
[35], 2006	35	20	1	71.6	5.6
[36], 2006	20	44	5.5	60	2.3
[37], 2007	268	40	20	64	5.2
[38], 2007	284	75	37	63.5	3.1
[39], 2007	909	125	49	67.3	4.9
[40], 2008	285	20	15	71.3	4.8
[41], 2008	91	45	22	53	5.6
[42], 2008	350	33	18	70.3	4.0
[43], 2008	1.2	250	117	28.4	0.24
[44], 2008	180	200	60	62.0	1.46
[45], 2009	1.44	50	25	49.2	0.12
[46], 2009	250	100	50	73.0	0.69
[47], 2009	9.9	50	25	58.2	0.30
[48], 2009	50	2,500	1,250	34.0	0.49
[49], 2009	385	125	62.5	78.0	0.47
[50], 2009	30.3	600	300	46.0	0.31
[51], 2009	92	100	50	67.6	0.47
[52], 2009	1,260	1,000	500	56.5	2.31
[53], 2009	55	500	233	52.8	0.33

Table 6.3. Examples of state-of-the-art pipelined AD converters without calibration.

Reference	Power (mW)	Sampling rate (MSps)	ERBW (MHz)	SNDR (dB)	FoM (pJ/conv.step)
[54], 2006	18	50	25	54.6	0.8
[55], 2008	230	100	46	72.2	0.75
[56], 2008	4.5	100	50	59.0	0.06
[57], 2009	12	50	25	58.4	0.35

FoM of $0.06 - 0.8$ pJ/Conversion-step. In view of that, the $0.05 - 0.2$ pJ/ Conversion-step of the digital algorithms is a relatively large amount when compared to a state-of-the-art overall performance. From this, it is concluded that in order to achieve state-of-the-art performance with a smart concept, not only the analog parts need improvement, but also the costs for the digital part have to be reduced.

3. High-speed high-resolution AD conversion

From the prior art discussed in the previous section, it becomes clear that AD converters with smart properties exist, but it is not yet evident whether they can improve the performance. In this section, a few key-factors will be defined

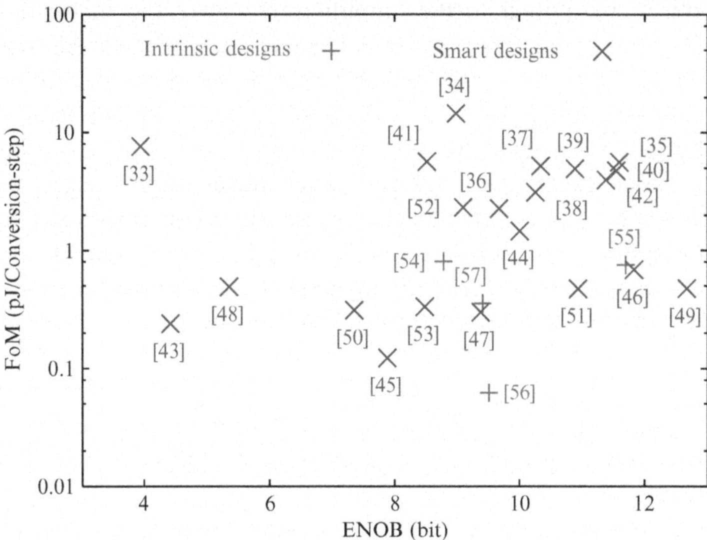

Figure 6.1. FoM as a function of the achieved ENOB for intrinsic and calibrated ADCs.

Table 6.4. Digital calibration performance.

Reference	Digital Calibration Power (mW)	$\mathrm{FoM}_{\mathrm{calibration}}$ (pJ/conv.step)
[34], 2006	1	0.20
[35], 2006	1.1	0.18
[38], 2007	11	0.18
[40], 2008	5	0.06
[53], 2009	8	0.05

that have the potential to enable high-performance AD converters targeting high-speed/high-resolution applications. Then, a smart approach will be used to improve the bottlenecks associated with these key-factors. The two key-factors that will be discussed are *open-loop circuitry* and *time-interleaving*.

3.1 Open-loop versus closed-loop

Amplifiers are used as a basic component in many different AD topologies, e.g. in the inter-stage gain block in a pipelined ADC, the buffer inside a sample-and-hold, the gain block in a cyclic ADC or the filter of a sigma-delta modulator. An important choice to be made is whether to implement the amplifier as a closed-loop structure or as an open-loop structure. In a closed-loop topology, an amplifier with a high open-loop DC gain is applied. Typically, the DC gain is chosen larger than 2^N, where N is the number of bits of the ADC.

Then, feedback is applied to the amplifier to realize the required transfer function. Because of the high open-loop DC gain and the feedback, this topology is accurate in terms of the realized transfer function. Alternatively, an open-loop amplifier could be used. As opposed to a closed-loop structure, this topology has a low DC-gain and does not apply feedback. The transfer function is determined directly by the open-loop characteristic of the amplifier, and is not improved by means of feedback. As a result, all the imperfections of the open-loop amplifier (e.g. non-linearity, mismatch, process spread, sensitivity to the environment) become directly apparent in the final performance, while the same effects are suppressed by the feedback in a closed-loop situation. From an accuracy point-of-view, a closed-loop design is preferable as the accuracy is well-defined by means of the feedback, while the accuracy of an open-loop design is not. However, when aiming for high-speed and low-power, the open-loop topology might be advantageous, as will be explained with an example next.

Figure 6.2 shows a simplified small-signal model of a single-stage amplifier, which could be part of either a closed-loop or an open-loop topology.[1] g_m and r_{out} model the transconductance and the output resistance of the stage, while c_{load} models the connected output load. Given a certain load and a certain speed, the closed-loop and open-loop solutions will be compared to each other with respect to power consumption. From the model, it follows that the DC gain A_0 is given by $A_0 = g_m \cdot r_{out}$ and the gain-bandwidth-product *GBP* by $GBP = g_m/c_{load}$. For a given load (c_{load}) and a given speed requirement (GBP), the g_m is known regardless of the implementation of the amplifier. The difference between closed-loop and open-loop is in the fact that for closed-loop, a large value is needed for A_0, while for open-loop, a small value is needed for A_0. As g_m is fixed, the difference in A_0 is realized by using either a high or a low value for r_{out}.

The power consumption of the single-stage amplifier is determined by the required g_m. For a single transistor, g_m is given by $g_m = \sqrt{2\beta I_D}$ [58], where I_D is the bias current of the transistor. From g_m, a minimum value for I_D can be determined, which yields a lower-bound for the power consumption of the amplifier. From this analysis, one might conclude that for a given speed,

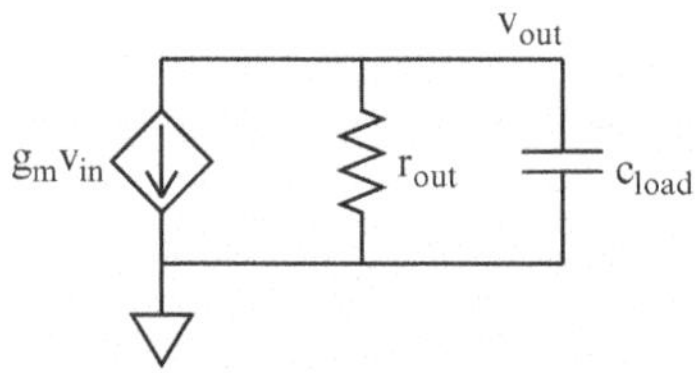

Figure 6.2. Small-signal model of a single-stage amplifier.

the closed-loop and open-loop solutions require the same amount of power. However, there are several reasons why in reality the power consumption of a closed-loop solution will be higher than that of an open-loop solution:

- First of all, the topology that implements the closed-loop amplifier has to be considered. For a telescopic topology [58], the power consumption remains indeed as explained before. However, the telescopic topology requires many stacked transistors, which is not feasible in modern IC technologies with a low-voltage supply. To reduce the stacking problem, a folded-cascode topology can be selected [58]. In this case, the input pair and the cascode devices require a separate bias current instead of reusing the same bias current. As a result, the power consumption of this topology will increase (typically with a factor of two) compared to the original expectation. For the open-loop case, there is less need for cascoded devices, as the gain-requirement is relaxed. Thus, low-voltage operation is feasible and the design can be implemented with a single bias current.

- A second reason for increased power consumption of a closed-loop solution is that, because of the high r_{out} requirement, transistors with a large L are needed. As a side-effect of increasing the transistor's L, the parasitic capacitances of these devices will also increase. In turn, this leads to either an increased power consumption or a reduced speed. In an open-loop solution, there is no need for transistors with a large L as the gain requirement is relaxed. As such, the parasitics in an open-loop solution can be smaller, resulting in a faster and more power-efficient design.

- A third issue with closed-loop designs is that, especially for higher resolutions, the required open-loop gain cannot be achieved with a standard low-voltage single-stage design [58]. Additional circuits (like gain-boosting circuits or a two-stage amplifier design) will be necessary to achieve the gain-requirement. As explained in [58], both these techniques result in a reduced speed and/or an increased power consumption.

From the previous analysis, it follows that the high DC gain of a closed-loop solution enables the high accuracy of such a design. However, at the same time, it is exactly the high DC gain that results in issues with respect to speed, power consumption and low-voltage operation. Table 6.5 summarizes the comparison between closed-loop and open-loop solutions. In view of a smart approach, the open-loop amplifier is selected for further implementation as it enables high-speed low-power operation with good portability to future technologies. Then, the smart concept will be applied to overcome the accuracy limitations. The implementation of the open-loop amplifier will be discussed in Chapter 7, while Chapter 8 deals with the smart solution for enhanced accuracy.

Table 6.5. Comparison of open-loop and closed-loop topologies.

	Closed-loop	Open-loop
Accuracy	+	–
Speed	–	+
Power	–	+
Low-voltage operation	–	+

3.2 Time-interleaving

A second key-factor in high-speed ADC design is the principle of time-interleaving [59]: instead of using one converter operating at a sampling rate f_s, a number of converters (p) is used in parallel in a time-interleaved mode of operation: each channel operates at f_s/p only, but the combination of converters acts as a single converter operating at f_s. By doing so, the overall speed (f_s) can be much higher than the speed of a single converter (f_s/p). Dependent on the situation, there are at least two reasons why time-interleaving can be a useful approach:

- The overall speed requirement might be so high, that it cannot be implemented by a single-channel converter due to fundamental or practical speed-limitations.
- From power-efficiency point-of-view, it can be advantageous to use several parallel converters with reduced speed instead of using a single high-speed converter.

Though it is beyond the scope of this work to investigate when time-interleaving is advantageous and when not, two simple examples will be shown to demonstrate that these effects are indeed relevant. As a first representative example in the context of AD conversion, a small-signal model of a basic analog block will be reviewed. The example considers a source follower, as it could be part of a T&H circuit. To demonstrate that time-interleaving can be advantageous in a much wider context, the second example considers the large-signal behavior of a digital circuit. For simplicity, a concatenation of two inverters will be reviewed.

Figure 6.3 shows a simple source follower, composed of a transistor with a bias current source and a capacitive load. The small-signal model is also shown, in which g_m, r_{out} and c_{par} model the transistor.

In this case, c_{par} models the parasitic capacitance of the transistor at the output node. The speed of this circuit is given by the pole:

$$f_{pole} = \frac{1 + g_m r_{out}}{2\pi r_{out} \cdot (c_{par} + c_{load})} \tag{6.3}$$

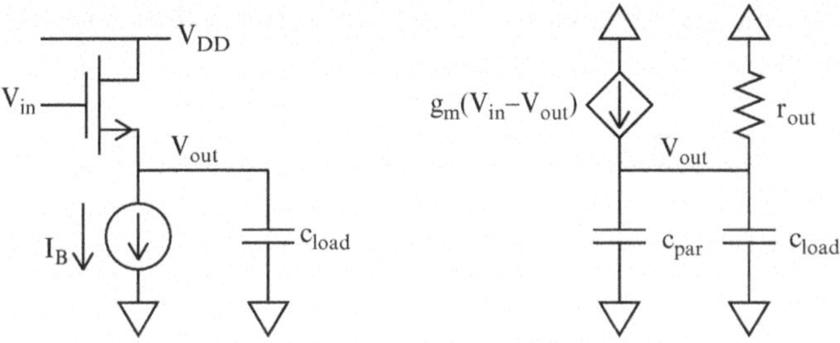

Figure 6.3. A source follower (left) and its small-signal model (right).

The speed of this circuit can be increased by scaling the current and the transistor size. When I_B and W/L are scaled with a factor k, g_m will also scale with k, while r_{out} scales with $1/k$. As the parasitic capacitance is approximately proportional to WL, c_{par} will scale with k when L is fixed. c_{load}, which is determined by an external load, does not scale as a function of k. The effect of scaling on power consumption and speed is considered next: as I_B scales with k, this implies that the power is scaled with k as well. The speed-scaling can be observed by the impact of k on the pole frequency:

$$f_{pole}(k) = \frac{1 + g_m r_{out}}{2\pi r_{out} \frac{1}{k} \cdot (k c_{par} + c_{load})} \tag{6.4}$$

Dependent on which capacitor is dominant (kc_{par} or c_{load}), the speed scales as follows:

- $kc_{par} \ll c_{load}$: f_{pole} scales with k.
- $kc_{par} \gg c_{load}$: f_{pole} remains constant.

In the first case, the speed scales with k while the power also scales with k. This is a fair trade-off as it maintains the same performance level when considering the FoM (6.1). On the other hand, in the second case, the speed is not increased even though the power is increased with a factor k; here the speed-limit of the circuit is reached. A general view of this process is shown in Fig. 6.4, where the speed is plotted as a function of the power. In the figure, roughly three situations can be distinguished:

I. For low-speed, low-power situations, scaling is effective as speed and power are linearly proportional; this corresponds to the situation where c_{load} is dominant.

II. When increasing the speed and power, at a certain time, the speed improvement is slowed down; this corresponds to the situation where both kc_{par} and c_{load} are important.

III. Ultimately, the speed cannot increase any further; this corresponds to the situation where kc_{par} is dominant.

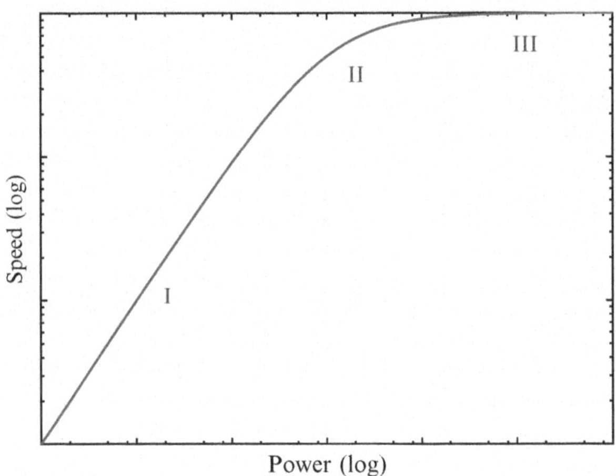

Figure 6.4. Trade-off between power and speed (arbitrary units).

With respect to the speed-power performance, only region I in Fig. 6.4 is efficient. As soon as a circuit operates in region II, it could be beneficial to apply time-interleaving. Consider a design with a certain power P and speed f_s in region II. Suppose time-interleaving is applied with (e.g.) two channels. Then, two new designs working at $f_s/2$ each are necessary. Because of the curvature of the speed-power relation in region II, each of these designs would consume less than $P/2$. Overall, these two channels still achieve the original speed f_s, but now at a power consumption which is less than the original P. As such, the 2-channel solution achieves a more power-efficient solution. However, it should also be noted that there will be additional circuit overhead when implementing a time-interleaved system: at the input, the signals should be split to the different channels, and at the output, the signals should be combined again. Time-interleaving will be advantageous only if the power reduction of the ADC core circuitry is higher than the added consumption of the overhead circuitry. In the last case, when a circuit is working in region III and it is necessary to further increase speed, again this could be done by means of time-interleaving.

From the above considerations, it can be concluded that in some situations time-interleaving improves the speed-power performance or it can increase the speed beyond a fundamental or practical speed limit. To show that this observation is not only valid for the presented analog block, a similar analysis is repeated for a digital block. Figure 6.5 shows a chain of two inverters, in which the wire capacitance is taken into account as it is often the dominant capacitance. A large-signal model is also shown, where the inverters are modeled by a controlled current source I_{out} and an input capacitance C_{in}. Note that I_{out} and C_{in} are large-signal parameters. When scaling is applied, similar to the previous example, again the power, I_{out} and C_{in} scale with k while c_{wire} remains constant. Effectively, the $\frac{dV}{dt}$ as a function of k for this example is given by:

$$\frac{dV}{dt} = \frac{kI_{out}}{kC_{in} + c_{wire}} \tag{6.5}$$

Note the similarity to the previous example: in this case speed-power scaling is efficient as long as c_{wire} dominates the overall capacitance, while the speed is limited when kC_{in} becomes dominant. Here, time-interleaving or parallelism becomes beneficial as soon as kC_{in} is not negligible compared to c_{wire}. In digital designs, parallelism is actually a well-known method to increase speed or efficiency.

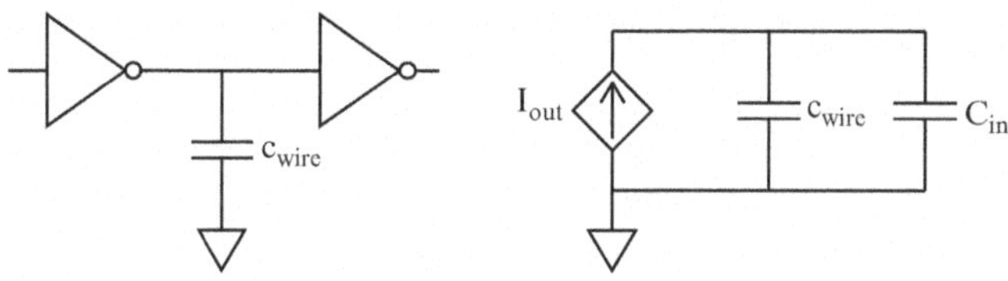

Figure 6.5. An inverter chain (left) and its large-signal model (right).

From the discussed examples, it can be understood that for specific situations, time-interleaving can improve the speed-power performance (by moving designs from region II to region I), or it enables the operation at a higher overall speed (by moving designs from region III to region I or II). In general, time-interleaving is a relevant option for high-speed AD converters. However, there is an issue with time-interleaved systems as they require proper matching between the individual channels. When mismatch is present, the overall performance can be degraded, as described in a.o. [60]. Because of the general usefulness of time-interleaving, a smart approach for the correction of channel-mismatch will be developed. The problem of channel-mismatch and the smart solution will be discussed in Chapter 9.

4. Smart calibration

In the previous section, two relevant factors that enable low-power, high-speed operation were determined: open-loop amplification and time-interleaving. However, open-loop structures have a limited accuracy, while time-interleaved structures are sensitive to channel mismatch. Because of that, a smart approach will be used to reduce the impact of these limitations. In this section, the global approach will be decided upon, while the actual implementations will be discussed in Chapters 8 and 9.

In line with the general view on smart conversion from Chapter 3, three components are required to reduce the effect of the imperfections of open-loop and time-interleaved structures:

1. **Detection** of the imperfections.
2. **Processing** of this information by means of an algorithm.
3. **Correction** of the effects of the imperfections.

Moreover, aiming for maximum performance, it is important that the additions for the smart solution (detection, processing and correction) improve the performance while using only limited resources; i.e. when adding the smart solution on top of the analog core, it should:

- Improve the overall accuracy.
- Reduce the speed as little as possible.
- Increase the power consumption as little as possible.

With these requirements in mind, a coarse decision can be made on how to implement the different parts of the smart solution. Several alternatives with respect to detection, processing and correction will be discussed next.

As will become clear in the following chapters, the imperfections of open-loop and time-interleaved structures are mainly dependent on random mismatch and process spread.[2] Though the statistics of random mismatch and process spread can be estimated a-priori, the actual values are known a-posteriori only. Because of that, on-chip detection is needed to acquire the a-posteriori information.

Then, for the calibration, several alternatives can be considered as visualized in Fig. 6.6. The main choice is between foreground and background calibration. With a foreground method, the normal operation of the ADC is interrupted during calibration. Therefore, a foreground method is especially suitable as a start-up calibration method: the calibration is performed once at the start-up of the circuit. Then, when the calibration is finished, the normal operation of the ADC starts. Alternatively, a foreground method can be used at certain intervals at runtime. This is suitable for applications in which the ADC is not used continuously, such that there are free time-slots available during which the calibration can take place. Obviously, when a foreground method would be used continuously at runtime, there would be no time left for normal operation. In case of a background method, the calibration can take place during normal operation of the ADC such that interruption is not necessary. Typically, a background method is used continuously at runtime. However, it could also be used at certain intervals or at start-up only; e.g. to reduce the power consumption of the method. Note that in principle, it is also possible to implement a combination of several alternatives, e.g. a foreground start-up calibration phase for coarse error correction together with a continuous background method for fine-tuning.

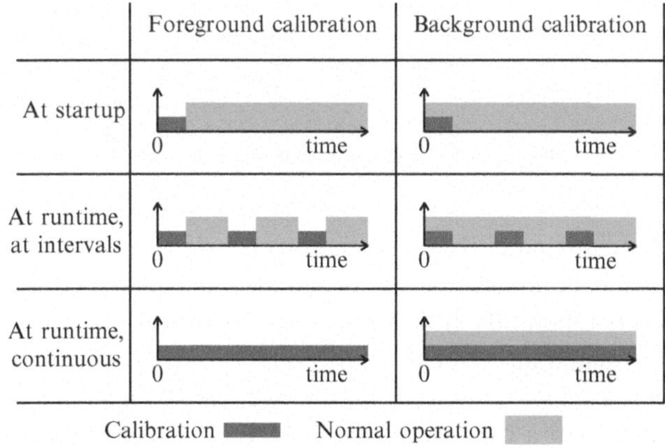

Figure 6.6. Alternatives for foreground and background calibration.

An important difference between foreground and background methods is that background methods can continuously track changes in the imperfections, while foreground methods can only operate at dedicated time-intervals. However, a foreground method has full control over the ADC during calibration, because of which it can acquire the required information in an efficient way in principal. On the other hand, the background method has to acquire information that is embedded within the normal data stream. As a result, background methods tend to have a higher complexity of the detection algorithm, they need a longer convergence time and they consume more power. As the goal of this work is a power-efficient solution, a foreground method is preferable. The fact that a foreground method cannot be used continuously is of less importance in this situation, as the relevant imperfections that are to be calibrated are constant at first order. Hence, a continuous detection algorithm is not required.

The last consideration is how to implement the actual correction of the detected imperfections. From the study in Section 2, it became clear that a digital post-correction is feasible, but that it can still use a considerable amount of power compared to the overall budget. As an alternative to digital correction, analog control is considered. Suppose the circuit in Fig. 6.7 is used as the input pair of an amplifier, and it is necessary to provide a means to control its gain.

This gain control can be achieved by adding a digital register that controls the bias current of the differential pair in discrete steps. As the gain of the amplifier is proportional to g_m, and $g_m = \sqrt{\beta I_B}$, tuning I_B results in a change of the gain. Because of the square-root relation, a tuning-range of (e.g.) $\pm 2\%$ for I_B results in a $\pm 1\%$ tuning-range of the gain. Nonetheless, as the gain

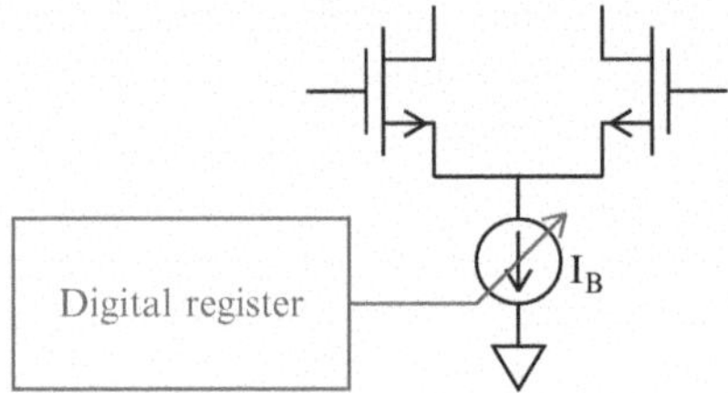

Figure 6.7. Example of analog control.

variation (or other mismatch-based variation) is relatively small, the variation of I_B is still small compared to the nominal value of I_B. Because of that, analog control has inherently little impact on the overall power consumption. Moreover, as mismatch has a mean-value equal to zero, it is equally likely that I_B is either decreased or increased; on the average (over many amplifiers, or over many channels in a time-interleaved ADC) I_B will remain equal to the nominal value. As the analog control is expected to be more power-efficient than the digital correction, analog control is selected for the actual correction of the relevant imperfections.

5. Conclusion

In this chapter, the general smart concept was narrowed down to a specific goal: to improve the performance of a high-speed, high-resolution ADC in terms of the speed/power/accuracy trade-off. From a literature study, it can be concluded that prior art in this direction exists, but most of the concepts are still lacking in absolute terms of performance. In order to improve upon the prior art, two key-factors were investigated that have the potential to enable high-performance AD converters for high-speed/high-resolution applications, namely:

- Open-loop circuitry.
- Time-interleaving.

Open-loop circuits can achieve higher speed and lower power consumption compared to closed-loop solutions. Besides, they are more suitable for new-generation low-voltage technologies. Currently, the limited accuracy of open-loop circuits is a main bottleneck. It was shown that the second factor, time-interleaving, enables a higher speed of operation and a higher power-efficiency for high-speed designs. However, time-interleaved circuits are prone to channel-mismatch errors. Though these two key-factors enable high performance in theory, their associated limitations (accuracy and channel-mismatch) need to be resolved. In order to reach that goal, the smart concept will be applied. To realize a power-efficient smart calibration method for the previously mentioned issues, the following approach was selected:

- Foreground calibration method, performed at start-up.
- Digital processing of the information.
- Analog correction of the imperfections.

For a proof-of-concept, instead of designing a complete ADC, an open-loop Track-and-Hold circuit will be designed as a test-case in Chapter 7. Even though it is not a complete ADC, it is sufficient to demonstrate the effectiveness of an open-loop design. Also, it enables the verification of the smart approach to resolve both the accuracy limitations of an open-loop T&H and the channel-mismatch of a time-interleaved T&H. These two smart solutions will be discussed in Chapters 8 and 9, respectively.

Notes

1. Note that multi-stage topologies are not considered in this work.
2. The effects of environmental changes or temperature gradients are not considered in this work.

Chapter 7

DESIGN OF AN OPEN-LOOP T&H CIRCUIT

This chapter presents the design and implementation of an open-loop T&H circuit. The current state-of-art in T&H design is discussed first. Then, after analyzing two alternative topologies for the implementation of the buffer, one specific topology is selected for the final implementation. Simulations and experimental results are shown and the performance is compared against existing solutions. In Chapters 8 and 9, smart calibration techniques will be added to the presented T&H design to further enhance the performance for either higher accuracy or time-interleaved operation. Parts of this chapter have been published previously in [61, 62].

1. Literature review

In most cases, T&H circuits are not published as a separate component, but as an integral part of a complete ADC. Because of that, limited information could be found on the performance of the T&H circuits itself. An overview of recent work on T&H circuits was made, considering experimentally verified CMOS implementations only. A summary is included in Tables 7.1 and 7.2, reviewing closed-loop and open-loop architectures, respectively. For the open-loop solutions, the used topology is indicated by either SF (source follower) or DP (differential pair). Note that the designs of [63] and [64] are realized by time-interleaving 16 channels; the information in the table is for a single channel.

Few publications report the achieved performance in terms of linearity and noise (expressed in SNDR or ENOB); in most cases, only the linearity (expressed in SFDR or THD) is given while noise is neglected. To accommodate for this, two different FoMs are used; one based on the ENOB,

P. Harpe et al., *Smart AD and DA Conversion*,
Analog Circuits and Signal Processing, DOI 10.1007/978-90-481-9042-3_7,

and one based on the SFDR:

$$FoM_{ENOB} = \frac{Power}{2^{ENOB} \cdot min(f_s, 2f_{in,max})}, \text{ with} \tag{7.1}$$

$$ENOB = \frac{SNDR - 1.76}{6.02}$$

$$FoM_{SFDR} = \frac{Power}{2^{(SFDR-1.76)/6.02} \cdot min(f_s, 2f_{in,max})} \tag{7.2}$$

Note that the SFDR-based FoM gives a lower-bound for the ENOB-based FoM, as the SNDR is upper-bounded by the SFDR.

Table 7.1. Overview of recent work on closed-loop T&H circuits in CMOS technology.

Reference	[65], 2001	[66], 2004	[67], 2005	[68], 2007	[69], 2008
Technology	0.5 μm	0.35 μm	0.18 μm	0.25 μm	0.35 μm
Power supply	1.2 V	3.3 V	3.3 V	0.5 V	3 V
Power consumption	1.2 mW	320 mW	75 mW	0.3 mW	26.4 mW
SFDR/THD	50 dB	65 dB	78 dB	–	66 dB
ENOB	–	–	–	9.3 bit	–
$min(f_s, 2f_{in,max})$	6 MHz	20 MHz	90 MHz	1 MHz	240 MHz
FoM-SFDR	774 fJ	11 pJ	128 fJ	–	67 fJ
FoM-ENOB	–	–	–	476 fJ	–

Table 7.2. Overview of recent work on open-loop T&H circuits in CMOS technology.

Reference	[70], 2001	[71], 2002	[63], 2004	[72], 2006	[64], 2007
Technology	0.35 μm	0.35 μm	0.12 μm	0.18 μm	0.13 μm
Power supply	3.3 V	3.3 V	1.2 V	1.8 V	1.6 V
Power consumption	70 mW	30 mW	2 mW	200 mW	4.6 mW
SFDR/THD	63 dB	35 dB	50 dB	28 dB	64 dB
ENOB	–	–	7.6 bit	–	7.7 bit
$min(f_s, 2f_{in,max})$	90 MHz	1 GHz	100 MHz	10 GHz	84 MHz
FoM-SFDR	674 fJ	650 fJ	77 fJ	975 fJ	42 fJ
FoM-ENOB	–	–	103 fJ	–	260 fJ
Topology	DP	SF	SF	SF	SF

When comparing the open-loop and closed-loop solutions with respect to speed, accuracy and FoM, the following can be observed, which corresponds to the analysis from Section 3.1:

- **Speed**: the open-loop solutions achieve a higher speed (90 MHz–10 GHz) compared to the closed-loop solutions ($\leq$240 MHz).
- **Accuracy**: on average, the linearity of the closed-loop solutions (50–78 dB) is better than the linearity of the open-loop solutions (28–63 dB).
- **FoM**: both closed-loop and open-loop solutions can achieve a FoM below 100 fJ.

2. Design goal

The main goal of this work is to verify the feasibility of smart techniques applied to important building blocks of AD converters, in this case the T&H circuit. For this purpose, it is required to have a chip-implementation of a T&H. Though not strictly necessary, it is preferable to have an implementation with a performance that is at least in the order of current state-of-the-art designs.

Though the T&H is considered as a stand-alone component in this work, it was actually implemented together with an ADC. As the ADC targets 8-bit performance at 500 MSps, the minimum requirement for the T&H was set at an SFDR of 60 dB and a sample rate of 500 MSps. Moreover, as the ADC was designed first, the requirements for signal-range, common-mode level and expected load were also defined beforehand. Aiming at an SFDR-based FoM of 50 fJ for the T&H, a first estimate of the power budget of the T&H was set to 25 mW. An overview of the design specifications is given in Table 7.3. It should be noted that even compared to today's state-of-the-art, the proposed design target is still a challenging goal: the best SFDR-based FoM reported in Section 1 equals 42 fJ, which is in the same order as the 50 fJ-target. In absolute terms of speed and linearity, the design goal is also at the limit of current solutions as illustrated in Fig. 7.1: only one existing solution achieves a marginally better speed-linearity product.

Supported by the theory presented in Chapter 6 and the study in the previous section, it was decided to implement the T&H as an open-loop structure. In the next section, the open-loop architecture is introduced.

Table 7.3. Design goal of this work.

Technology	0.18 μm
Power supply V_{DD}	1.8 V
Signal range $V_{in,pp}$	1.0 V
Common mode voltage V_{CM}	1.1 V
Load	500 fF
Power consumption	≤25 mW
SFDR	≥60 dB
$f_s, 2f_{in,max}$	500 MHz
FoM-SFDR	≤50 fJ

3. T&H architecture

A general view of a differential open-loop T&H circuit is given in Fig. 7.2. The analog time-continuous input signal is sampled onto the sampling capacitors by means of switches. The switches are controlled by an externally applied clock signal through a switch driver. An open-loop output buffer is used to drive the load (the ADC) without affecting the sampled value at the capacitors.

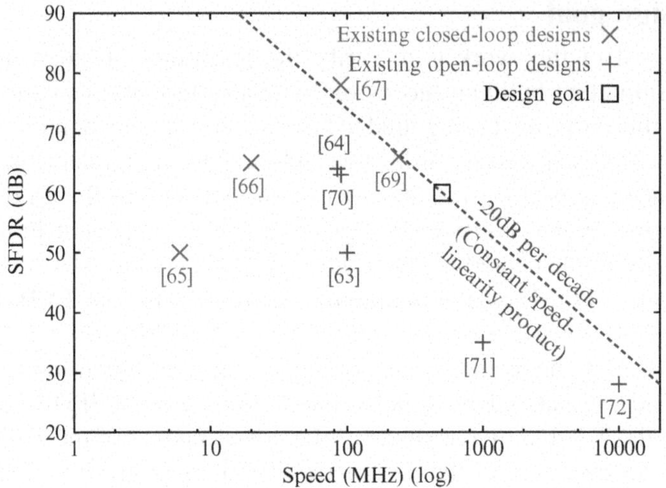

Figure 7.1. Speed and linearity of existing T&H circuits and the chosen design target.

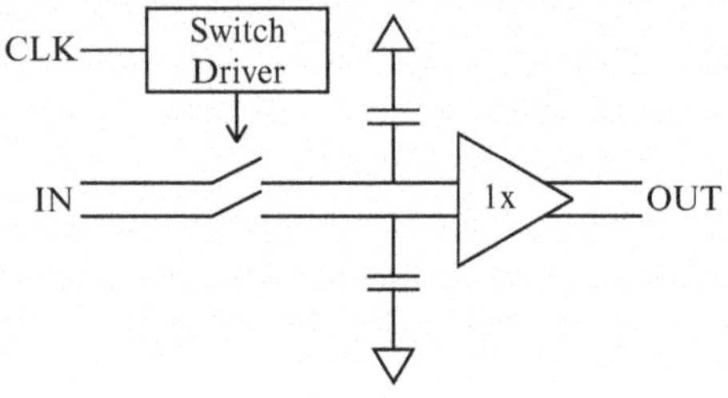

Figure 7.2. Open-loop Track&Hold architecture.

It should be noted that this architecture is actually composed of two open-loop structures: the first one is the sampling structure itself (switches, switch drivers and capacitors), and the second one is the output buffer which is to be implemented as an open-loop circuit as well. First, the sampling structure will be discussed in Section 4. Then, in Section 5, two alternative implementations for the open-loop buffer will be analyzed. The implementation of the complete T&H will be discussed in Section 6, followed by experimental results in 7.

4. Sampling core architecture

The actual core of the T&H circuit is the sampling circuit, composed of the sampling capacitors, switches and switch drivers (Fig. 7.2). The size of the sampling capacitors was set to 200 fF, such that for a full-scale input sine ($1V_{pp}$) an SNR of around 64 dB is achieved. The switches use the boot-strapping technique presented by [73] to achieve both high speed and high linearity. Using this technique, the actual switch can be implemented with a single NMOS device. High speed is obtained by driving the switch with a high overdrive voltage $V_{gs} = V_{DD} = 1.8$ V because of which a small transistor ($\frac{W}{L} = \frac{5\ \mu\text{m}}{0.18\ \mu\text{m}}$) can be used as switch, which in turn reduces the parasitic capacitance. As a result of the high overdrive voltage, a small on-resistance is still achieved. Next to that, as the bootstrapping technique generates a constant V_{gs} voltage, independent on the input signal V_s at the switches' source, high linearity is achieved as well. The implementation of the switch driver is shown in Fig. 7.3, which is identical to the design described by [73]: the capacitors are pre-charged to act as an internal 1.8 V battery. When CLK is low, the gate of the sampling switch is connected to ground to open the switch. When CLK is high, the 1.8 V battery will be connected between the source and the gate of the sampling switch, such that $V_{gs} = V_{DD} = 1.8$ V and the switch will be turned on.

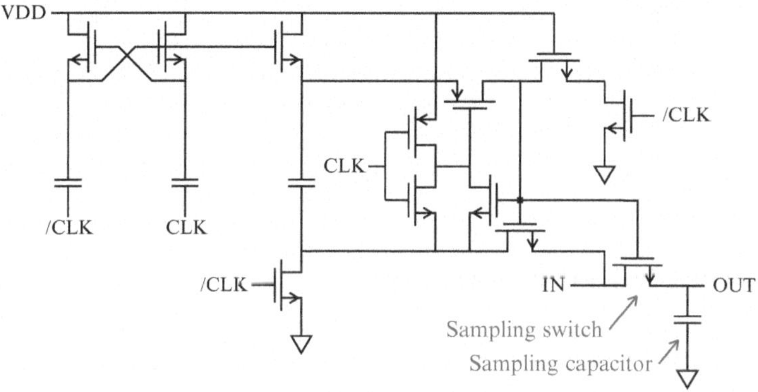

Figure 7.3. Bootstrapping technique applied to the sampling switch.

Transistor-level simulations were performed on the sampling core (excluding the output buffer), with the implemented capacitors, switches and switch drivers. While a slow input ramp signal was applied to the input, the sampling core was sampling at a rate of 500 MSps. The sampled data points were stored for processing. After subtraction of the best-fit line, the distortion introduced by the sampling core remains. This distortion curve is plotted in Fig. 7.4. The maximum deviation is only $V_{err,max} = 32\ \mu$V. This performance can be expressed in terms of effective-number-of-bits (ENOB) by equating this error with 0.5 LSB of the full-scale range ($\pm V_{fs}$):

$$ENOB = \log_2 \frac{V_{fs}}{V_{err,max}}, \tag{7.3}$$

resulting in an equivalent accuracy of $ENOB = 13.9$ bit, which is sufficient for the design goal of about 10 bit linearity. The simulated power consumption equals 0.25 mW, which is negligible compared to the overall T&H consumption. The performance of the switch-driver was not further optimized as the achieved linearity and power consumption are abundantly better than the expected overall performance.

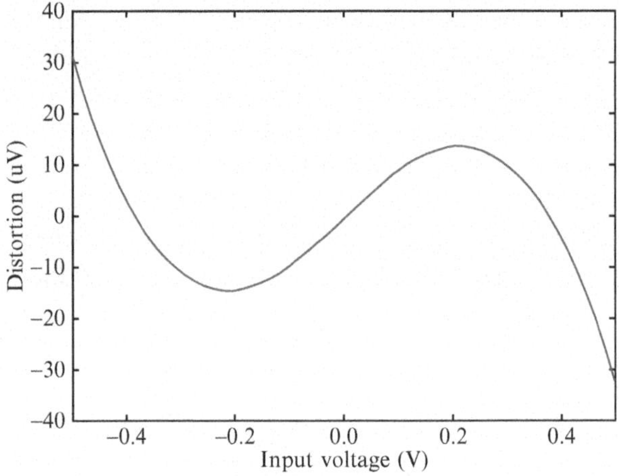

Figure 7.4. Simulated distortion of the sampling core, operating at 500 MSps.

5. Output buffer architecture

Based on the study in Section 1, two possible solutions for the required open-loop buffer will be reviewed here. The first solution is based on a source-follower, the second on a differential pair. After a brief introduction of the two alternatives, a comparison is made with respect to speed, power consumption, accuracy, mismatch sensitivity, controllability and power supply requirements.

5.1 Source follower

A first option for an open-loop unity-gain buffer is a source follower. A pseudo-differential source follower using NMOS transistors is illustrated in Fig. 7.5. Transistors M_1 and M_2 are biased with a constant V_{GS} (equal to V_B), such that they generate a constant current of $0.5I_B$ each. Therefore, the total current consumption equals I_B. The differential input voltage is applied at the gates of transistors M_3 and M_4. As a first order approximation, a constant current $0.5I_B$ flows through each of these transistors, resulting in a constant V_{gs}. As such, the source potentials will track the gate potentials, generating a differential output voltage equal to the differential input voltage. At the same time, because of the V_{gs} voltage drop across transistors M_3 and M_4, the common-mode level at the output will be lower than the common-mode level at the input. In situations where a common-mode level-shift is undesirable, a second level-shift is necessary to compensate for it. In this section, it is assumed that there is no specific constraint on the common-mode level, such that a second level-shift can be omitted.

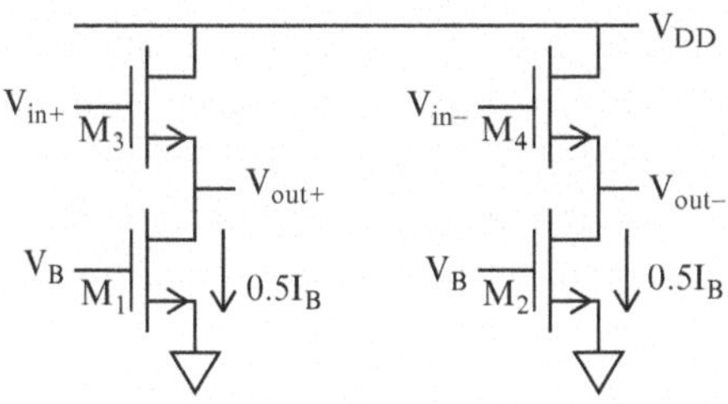

Figure 7.5. Pseudo-differential source follower.

5.2 Differential pair

An alternative for the output buffer is a differential pair with resistive load as shown in Fig. 7.6. Transistor M_1 is used as current source, setting the overall current to I_B as before. The differential pair converts the differential input voltage to a differential current. Then, this current is converted to an output voltage

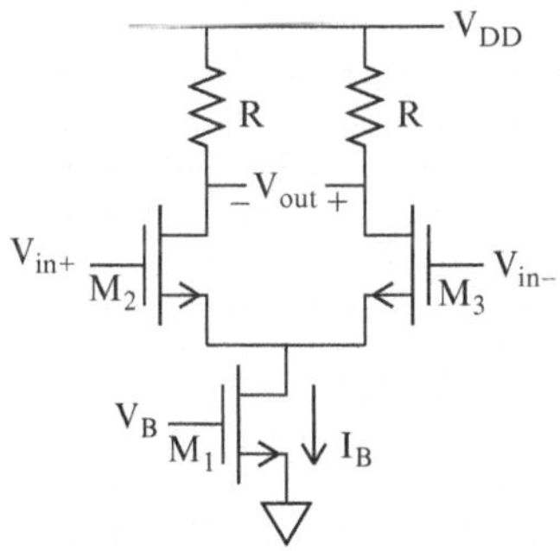

Figure 7.6. Differential pair.

by means of the resistors. In contrast to the situation with the source follower, no inherent level-shift is present in this case. As long as proper biasing of all devices can be maintained, the output common-mode can be set independent of the input common-mode.

5.3 Comparison of the two architectures

In this section, a comparison between the two proposed architectures is made.

Power consumption and speed

Both architectures introduce two time-constants which will limit the speed of the T&H. The first time-constant is related to the output resistance of the preceding stage combined with the input capacitance of the buffer. Note that the input capacitance of the buffer is placed in parallel to the sampling capacitors (Fig. 7.2). As for higher accuracies, the sampling capacitance is normally dominant over the buffer capacitance due to noise requirements, this time-constant shows a minor dependency on the buffer design. The second time-constant is related to the output resistance (r_{out}) of the output buffer combined with the input capacitance of the ADC (c_{load}). As c_{load} is assumed to be constant for both alternatives, the time-constant is dependent only on r_{out}. The output resistance of both buffers can be estimated using the following transistor relations:

$$I_d = \frac{1}{2}\beta(V_{gs} - V_{th})^2 \text{ and } g_m = \sqrt{2\beta I_D}. \tag{7.4}$$

By means of a small-signal analysis of Fig. 7.5, the single-ended output resistance of the source follower can be derived:

$$r_{out} = \frac{1}{g_m} = \frac{1}{\sqrt{\beta I_B}}, \tag{7.5}$$

which results in a relation between speed ($\tau = r_{out}c_{load}$), power consumption ($P = I_B V_{DD}$) and transistor dimensions ($\beta = \beta_\square \frac{W}{L}$):

$$\tau \propto \frac{1}{\sqrt{\frac{W}{L}P}}. \tag{7.6}$$

In case of the differential pair from Fig. 7.6, one can derive that its gain is given by $A = g_m r_{out}$. As the output buffer is designed to achieve unity-gain ($A = 1$), this implies that $g_m r_{out} = 1$. Therefore relations (7.5) and (7.6) are also valid for this architecture. As a result, with respect to the trade-off between power consumption and speed, the circuits perform identical.

Accuracy, mismatch sensitivity and controllability

Based on the simplified transistor equations (relation (7.4)), a source follower is perfectly linear and achieves unity-gain independent of the exact current I_B or the transistor dimensions. This means that, under assumption of this model, the linearity of the buffer is not adversely affected by mismatch of the components, process spread or a deviation of the biasing conditions. On the other hand, this also implies that the designer has little control over the realized gain and linearity, as these properties are relatively insensitive to the main design parameters, namely bias current and transistor dimensions. In practice, the accuracy of the source follower is limited by the effects that are not included in the approximation from (7.4). Secondary effects (e.g. channel-length modulation and body-effect [58]) introduce both signal distortion and gain drop. The severeness of these secondary effects, and therefore the accuracy of a source follower circuit, is dependent on the design and the used technology.

As opposed to the source follower, a differential pair is always inherently a non-linear circuit, even when using the simplified transistor equations (7.4). Also, the gain is not approximating unity by default, but equals $A = g_m r_{out}$, which can be chosen by the designer. As gain and linearity are dependent on the first order model, they can be well controlled by the designer using the main design parameters like bias current, transistor dimensions and resistor values. At the same time, the relatively high sensitivity of the performance to the first order effects implies that the actual performance will be sensitive to mismatch of components as well.

In certain situations, the intrinsic accuracy of an open-loop circuit might not be enough, and additional correction mechanisms will be necessary to enhance the performance. One solution is to perform the correction in the analog domain, by tuning a certain set of parameters of the circuit. For example, a gain-correction can be implemented by tuning the bias current in the differential pair of Fig. 7.6. A source follower is not very suitable for analog correction due to the low sensitivity of the performance to the parameters of the circuit. In other words, the *controllability* of the differential pair after production is higher than the controllability of the source follower.

Concluding, the performance of the source follower is mismatch unsensitive but also allows little freedom to the designer and little control after production. The performance of the differential pair is more sensitive to mismatch but gives more freedom to the designer for optimization and the performance can be controlled better after production. Table 7.4 summarizes the expectations which will be verified in Section 6, where a design example will be considered.

Power supply requirements and portability

For the portability to new CMOS technologies, the minimum required power supply is an important criterium. For the circuit in Fig. 7.5, both transistors should operate in pinch-off mode ($V_{ds} \geq V_{gs} - V_{th}$) for the full range of

Table 7.4. Comparison of accuracy, mismatch sensitivity and controllability.

	Source follower	**Differential pair**
Gain **Linearity**	Determined by secondary effects	Determined by first order effects
Mismatch sensitivity **Design freedom** **Controllability**	Low	High

input signals. It is assumed that all transistors need a certain overdrive voltage ($V_{gs} \geq V_{th} + V_{ov}$) to operate properly, that the maximum differential input signal equals $\pm V_{fs}$, and that all signal levels should remain between 0 V and V_{DD}. In that case, a lower bound for the power supply can be derived:

$$V_{DD,min} = 2V_{ov} + V_{th} + V_{fs} \tag{7.7}$$

Using the same assumptions for the differential pair yields as lower bound for the power supply:

$$V_{DD,min} = 2V_{ov} + 2V_{fs}, \tag{7.8}$$

which means that the source follower is preferable when $V_{th} < V_{fs}$, and the differential pair is preferable otherwise. In reality, V_{th} will be in the order of 0.2 V up to 0.5 V (dependent on the technology), while typical values of V_{fs} are also chosen in this range (note that $V_{pp} = 2V_{fs}$). Therefore, in most cases this difference will be small. In general, both topologies will remain applicable with scaling of technology as V_{fs} can be adjusted by the designer and the remaining term ($2V_{ov} + V_{th}$ in case of the source follower) is small enough to fit into V_{DD}.

Conclusion

From the above, one can conclude that the two alternatives show identical performance with respect to speed and power consumption. Also, both architectures are suitable for low-voltage operation, allowing portability to future technology generations. The main difference between the architectures is the accuracy. Most importantly, the non-linearity of a source follower is determined by secondary effects, whereas the non-linearity of a differential pair is determined by primary effects. As such, it can be expected that the non-linearity of a differential pair is its most important drawback compared to a source follower. In the following section, the non-linearity of a differential pair will be studied in detail and several solutions are proposed to enhance the performance. Then, in Section 6, the enhanced differential pair will be compared against the source follower using a design example.

5.4 Enhanced differential pair

The non-linear transfer characteristic of a basic differential pair with resistive load (Fig. 7.7) can be derived using the relation $I_d = \frac{1}{2}\beta(V_{gs} - V_{th})^2$ (with $\beta = \mu_n C_{ox}\frac{W}{L}$), yielding:

$$V_{out} = V_{in}\sqrt{\beta I_B - (\frac{1}{2}\beta V_{in})^2} \cdot R\,, \tag{7.9}$$

which can be approximated by a Taylor series:

$$V_{out} \approx a_1 V_{in} + a_3 V_{in}^3 + a_5 V_{in}^5 + \ldots\,, \text{ with:} \tag{7.10}$$

$$\begin{cases} a_1 &= \sqrt{\beta I_B} R \\ a_3 &= -\dfrac{\beta^2 R}{8\sqrt{\beta I_B}} \\ a_5 &= -\dfrac{\beta^3 R}{128\sqrt{\beta I_B} I_B} \end{cases}$$

As the third-order distortion component is dominant, the higher order terms will be neglected from now on.

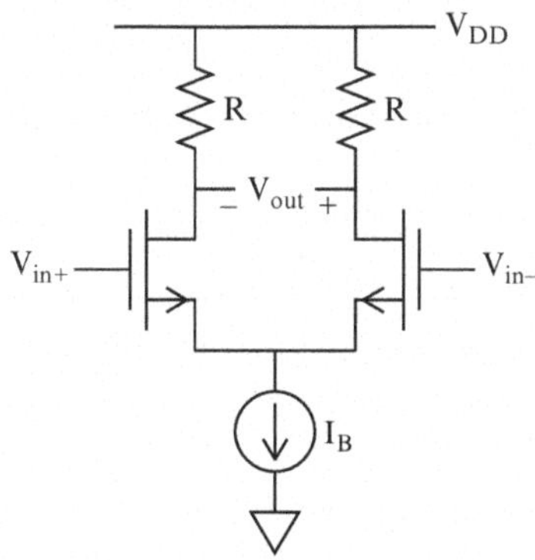

Figure 7.7. Basic differential pair with resistive load.

Because of the application, a_1 in (7.10) will be designed for unity-gain: $a_1 = 1$. Moreover, the relation $\frac{1}{2} I_B R = V_{DD} - V_{CM}$ should hold, where V_{CM} is the common-mode output voltage of the buffer. With these constraints, it follows that:

$$a_1 = \sqrt{\beta I_B} R = 1 \Rightarrow \beta = \frac{1}{I_B R^2} \tag{7.11}$$

Such that the third-order distortion term a_3 can be rewritten as:

$$a_3 = -\frac{\beta^2 R}{8\sqrt{\beta I_B}} = -\frac{1}{8 I_B^2 R^2} = -\frac{1}{32(V_{DD} - V_{CM})^2} \tag{7.12}$$

As a result, (7.10) simplifies into:

$$V_{out} \approx V_{in} + a_3 V_{in}^3 \text{ , with: } a_3 = -\frac{1}{32(V_{DD} - V_{CM})^2} \tag{7.13}$$

The third-order term $a_3 V_{in}^3$ expresses the dominant non-linearity component of the differential pair, used as unity-gain buffer. The maximum deviation $V_{err,max}$ with respect to the linear term is achieved for a full scale input signal $V_{in} = V_{fs}$:

$$V_{err,max} = |a_3 V_{in}^3| = \frac{V_{fs}^3}{32(V_{DD} - V_{CM})^2} \ . \tag{7.14}$$

Note that this result implies that when the voltage levels of the buffer are given, the linearity is also known. It is not possible to improve the linearity by e.g. altering the bias current or the transistor dimensions. For the given design goal (described before in Table 7.3) $V_{DD} = 1.8$ V, $V_{CM} = 1.1$ V and $V_{fs} = 0.5$ V, which results in $V_{err,max} = 8$ mV. In terms of effective-bits, this deviation translates to an ENOB estimation of:

$$ENOB = \log_2 \frac{V_{fs}}{V_{err,max}}, \tag{7.15}$$

yielding for the basic differential pair a maximum accuracy:

$$ENOB = \log_2 \frac{32(V_{DD} - V_{CM})^2}{V_{fs}^2} = 6.0 \text{ bit}. \tag{7.16}$$

The circuit of Fig. 7.7 was simulated with Cadence using the UMC CMOS 0.18 μm technology. The resistors were set to $R = 175\ \Omega$ to achieve a pole-frequency of almost 1 GHz with a load of 500 fF, the tail current source was set to 8 mA to ensure the correct V_{CM} of 1.1 V. Then, the transistor dimensions were adjusted in order to achieve a gain of one, resulting in $\frac{W}{L} = \frac{11.64\ \mu\text{m}}{0.18\ \mu\text{m}}$. From a DC analysis, the parameters of the polynomial function (7.10) can be estimated, yielding $a_1 = 1.00$ and $a_3 = -0.05$. From this, it follows that $V_{err,max} = 6.3$ mV and $ENOB = 6.3$ bit, which is close to the expected value of 6.0 bit.

In order to improve the linearity beyond the 6-bit level, two linearization techniques (known from a.o. [74]) will be applied: source degeneration and cross coupling. The first step is to add a source-degeneration resistor to the differential pair (Fig. 7.8). The resistor linearizes the transfer characteristic but it also reduces the gain of the differential pair. To maintain unity-gain, the transistors' $\frac{W}{L}$ has to increase compared to the basic differential pair.

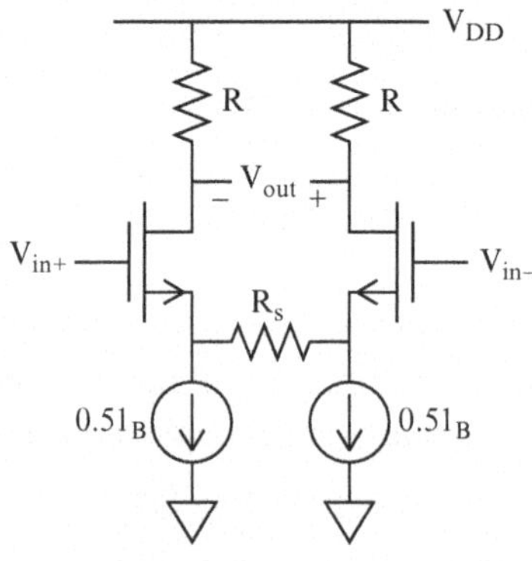

Figure 7.8. Differential pair with resistive source degeneration.

As a second modification, cross-coupling has been applied. Figure 7.9 shows the final implementation: a first differential pair (composed of transistors M_1, M_2, M_3, M_4 and source degeneration resistor R_{s1}) is connected as usual. A second differential pair (composed of transistors M_5, M_6, M_7, M_8 and source degeneration resistor R_{s2}) is cross-coupled to the first pair, in other words: the input is in parallel to the input of the first pair, but the output is reversed compared to the first pair. As before, the transfer function of each individual differential pair can be expressed as a polynomial function. Only

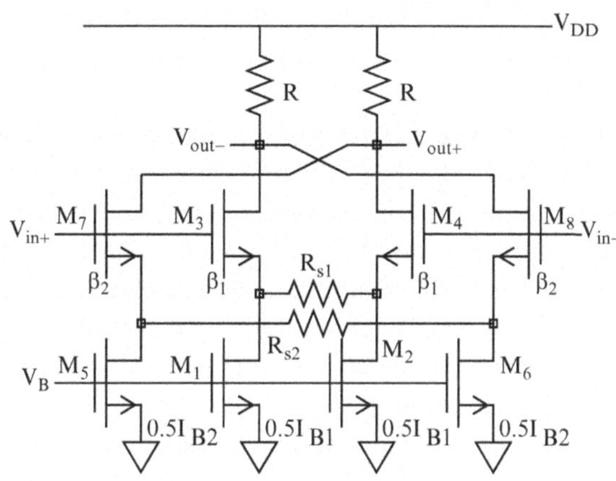

Figure 7.9. Cross-coupled differential pair with resistive source degeneration.

third-order and fifth-order distortion are taken into account, as they are dominant over the remaining components. The transfer function of the first pair can be written as:

$$V_{out,1} = a_1 V_{in} + a_3 V_{in}^3 + a_5 V_{in}^5 \, , \tag{7.17}$$

where coefficients a_1, a_3 and a_5 describe the polynomial function. a_1, a_3 and a_5 are dependent on the actual design parameters of the differential pair (the bias current, transistor dimensions and values of the resistors). Similarly, the transfer function of the second pair can be written as:

$$V_{out,2} = b_1 V_{in} + b_3 V_{in}^3 + b_5 V_{in}^5 \, , \tag{7.18}$$

where coefficients b_1, b_3 and b_5 describe a similar function as (7.17) but taking into account the device parameters of the second pair. The overall transfer function is the difference between (7.17) and (7.18), as the outputs of both pairs are inverted with respect to each other:

$$V_{out} = c_1 V_{in} + c_3 V_{in}^3 + c_5 V_{in}^5 \text{ , with:} \tag{7.19}$$

$$\begin{cases} c_1 &= a_1 - b_1 \\ c_3 &= a_3 - b_3 \\ c_5 &= a_5 - b_5 \end{cases}$$

By optimizing the parameters of both pairs (bias currents, transistor dimensions and source resistors), the circuit can be designed such that unity-gain is achieved ($a_1 - b_1 = 1$), while the dominant third-order distortion term is being cancelled: $a_3 - b_3 = 0$.

For simplicity of analysis, the polynomial model of a differential pair without source degeneration was taken to illustrate the principle of compensation by means of cross-coupling. The result of this analysis was used as a starting point for the circuit optimization of the cross-coupled pair with source degeneration. It should be noted that all calculations are based on the simple quadratic relation ($I_d = \frac{1}{2}\beta(V_{gs} - V_{th})^2$). Because of that, simulations are performed for the final optimization of the compensation scheme.

For a pair without degeneration, relation (7.10) holds. Aiming for unity gain and compensated third-order compensation, a total of six equations needs to be satisfied:

$$\begin{matrix} a_1 &= \sqrt{\beta_1 I_{B1}} R & b_1 &= \sqrt{\beta_2 I_{B2}} R & a_1 - b_1 &= 1 \\ a_3 &= -\frac{\beta_1^2 R}{8\sqrt{\beta_1 I_{B1}}} & b_3 &= -\frac{\beta_2^2 R}{8\sqrt{\beta_2 I_{B2}}} & a_3 - b_3 &= 0 \end{matrix} \tag{7.20}$$

For a given ratio m between the two transistor sizes $\beta_1 = m\beta_2$, there is only one ratio between the bias currents that meets the distortion requirements (7.20):

$$a_3 - b_3 = 0 \text{ and } \beta_1 = m\beta_2 \qquad \Rightarrow \qquad I_{B1} = m^3 I_{B2} \tag{7.21}$$

With this solution, the coefficients c_i of the cross-coupled pair can be calculated and expressed as a function of the coefficients of the main pair a_i:

$$\begin{cases} c_1 &= a_1(1 - m^{-2}) \\ c_3 &= 0 \\ c_5 &= a_5(1 - m^2) \end{cases} \tag{7.22}$$

Table 7.5 lists the gain and remaining dominant distortion component for different ratios m between the two differential pairs. Though not strictly necessary, only integer values are considered for m, as it is more convenient for the transistor-level implementation. Obviously, for $m = 1$ the two pairs are equal: all terms, including the linear term, will be equal to zero. As the output of the second pair is subtracted from the main pair, it will reduce the gain: the larger the ratio m between the pairs, the less signal will be subtracted and the closer the gain c_1 will be with respect to the original gain a_1: for $m = 2$, a gain-loss of 25% is introduced, but e.g. for $m = 4$, the gain-loss is reduced to 6%. Note that for all m (except for $m = 1$), a gain of one can be realized after cross-coupling by increasing the gain a_1 of the original pair. Even though a gain-loss might not be that critical from a function point of view, indirectly it does degrade the power-efficiency of the circuit. Because of that, a smaller gain-loss is preferable.

Table 7.5. Gain and fifth-order distortion for various cross-coupled pairs.

Ratio	**Gain**	**fifth-order distortion**
m	c_1	c_5
1	0	0
2	$3/4 \cdot a_1$	$-3a_5$
3	$8/9 \cdot a_1$	$-8a_5$
4	$15/16 \cdot a_1$	$-15a_5$

With respect to gain, it is best to maximize m as it reduces the gain-loss. On the other hand, increasing m also increases the amplitude of the remaining distortion component c_5 as illustrated in Table 7.5. On top of that one should note that for higher m, the current densities of the two pairs deviate more: as $\beta_1 = m\beta_2$ and $I_{B1} = m^3 I_{B2}$, the current density will differ by a factor of m^2. As such, the assumption that both pairs can be modeled by the same polynomial function (7.10) will be less plausible for higher m. As a compromise, m was set to 3 to limit the gain-loss to about 10%. With this starting point, the source degeneration resistors were optimized for linearity. The parameters in the final design are summarized in Table 7.6, while the simulated performance is visualized in Fig. 7.10. This figure shows the DC transfer function of the differential pair after subtraction of the linear term, such that only the distortion components remain. Also, a polynomial approximation is shown, which is composed of a third-order and a fifth-order term. As the polynomial approximation matches the simulated curve closely, it can be concluded that the third-order and fifth-order distortion are still dominating the overall linearity. The simulation results reveal that the maximum deviation was reduced to $V_{err,max} = 0.57$ mV, resulting in $ENOB = 9.8$ bit, which suits the 60 dB SFDR-goal. More precise linearity simulations will be presented later in this chapter, when the sampling core and output buffer will be simulated together.

Finally, noise simulations were performed to estimate the SNR of the output buffer. Results are shown in Fig. 7.11, yielding a noise power of $P_n = 50 \cdot 10^{-9}$ V^2. With a signal power of $P_s = \frac{1}{2}V_{fs}^2 = 0.125$ V^2, this gives an SNR of 64 dB.

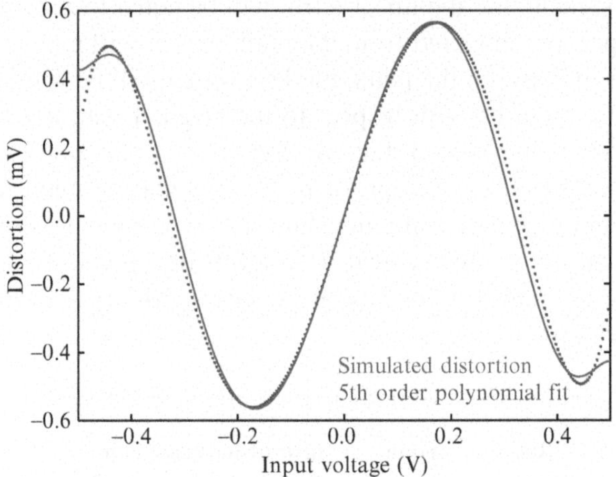

Figure 7.10. Simulated distortion of the output buffer.

Table 7.6. Parameters of the output buffer.

Parameter	Value	Parameter	Value
I_{B1}	7.8 mA	I_{B2}	0.3 mA
R_{s1}	307 Ω	R_{s2}	1.7 kΩ
W/L M_1,M_2	45.0 μm/0.18 μm	W/L M_5,M_6	1.2 μm/0.18 μm
W/L M_3,M_4	39.6 μm/0.18 μm	W/L M_7,M_8	13.2 μm/0.18 μm
R	167.5Ω		

5.5 Conclusion

In this section, two alternative architectures for the buffer were reviewed: a source-follower and a cross-coupled differential pair with source degeneration. Both architectures can meet the requirements of the design goal set previously. Therefore, both alternatives will be considered in the following section, where the complete T&H will be designed on transistor-level.

6. T&H design

In this section, two transistor-level implementations of an open-loop T&H circuit will be presented, both aiming for the design goal as proposed in Section 2. The output buffer was implemented with either a source follower or an enhanced differential pair, while the sampling core (see Section 4) remained identical in both cases. For a fair comparison, the current consumption of the output buffer was set to $I_B = 8$ mA in both cases. As such, the achievable gain, linearity and bandwidth of both output buffers can be compared for constant sampling speed and constant power consumption.

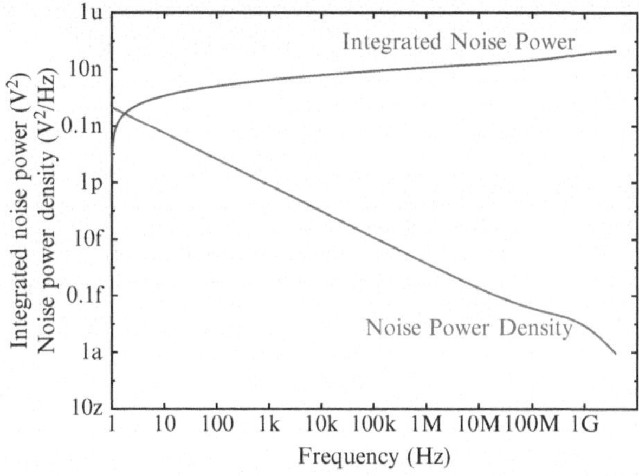

Figure 7.11. Simulated noise power of the output buffer.

6.1 Source follower based architecture

The source follower was implemented according to Fig. 7.5. As the bias-current I_B was fixed, the only remaining design-parameter is the size of the transistors. Minimum-length transistors ($L = 0.18$ μm) were chosen to minimize the parasitic capacitances. For the width, two options will be verified: a small width ($W = 40$ μm, the same value as used in the differential pair) and a large width ($W = 320$ μm). These two widths were selected as they give a clear impression of the effect of the size of the transistors on the performance of the source follower, as shown later by the simulation results.

6.2 Differential pair based architecture

For the differential pair, the enhanced structure from Section 5.4 was used. Based on the selected bias current and the sizes of the transistors composing the main differential pair ($W = 40$ μm, $L = 0.18$ μm), the remaining components were optimized. The precise design approach can be found in Section 5.4.

6.3 Simulation results

Transistor-level simulations were carried out in the time-domain to analyze the properties of the presented circuits. In all cases, a 500 MHz sampling clock was applied to the T&H, while sinusoids of various frequencies were applied as an input signal. Four T&H configurations were tested (summarized in Table 7.7): two configurations based on a source follower, one on a differential pair and one on an ideal buffer with zero input capacitance, zero output resistance and unity-gain. The ideal configuration determines the performance of the sampling structure (switches, switch-drivers and sampling capacitors), giving an upper-limit of the performance of the whole T&H.

Table 7.7. Simulated configurations.

Configuration	Description
SF1	Source Follower with $W = 40$ μm, $L = 0.18$ μm
SF2	Source Follower with $W = 320$ μm, $L = 0.18$ μm
DP	Differential pair with $W = 40$ μm, $L = 0.18$ μm
ID	Ideal output buffer

The frequency response of the complete T&H for input frequencies between 3 MHz and 2 GHz is given in Fig. 7.12. It can be seen that mainly the source follower structures deviate from unity-gain. Increasing the transistors' $\frac{W}{L}$-ratio reduces the loss, but is not effective because of the low sensitivity. Next to that, the source follower architectures show a peak in their response due to parasitic coupling from input to output. In case of the differential pair, the response behaves more smoothly. Nonetheless, all responses remain flat within ±0.1 dB up to the Nyquist frequency (250 MHz).

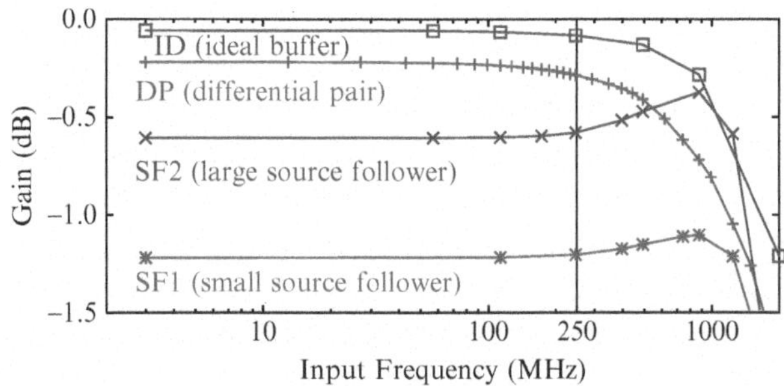

Figure 7.12. Response as a function of the input frequency.

The static linearity (Figs. 7.13 and 7.14) is limited by the output buffer (62 dB for the DP configuration, 68 dB for SF1 and 77 dB for SF2), as the linearity of the sampling structure (86 dB) is far better. The dynamic linearity was determined by means of the achieved SFDR (Spurious-Free-Dynamic-Range) as a function of f_{in} (Fig. 7.15). Clearly, the selection of the transistor size in the source follower has a dominant effect on its linearity: large transistors (SF2 compared to SF1) improve the static performance, but they degrade the dynamic performance for higher input frequencies. One can also observe that for high input frequencies, the linearity of the source-follower remains the bottleneck for the overall linearity. Opposed to this, the linearity of the differential pair degrades more slowly as a function of the applied frequency, such that for high frequencies the sampling core instead of the output buffer becomes the dominant source of distortion. Even though the source follower is superior for static linearity, the differential pair can be superior for high input frequencies (in this case far beyond Nyquist). Despite the differences, both architectures achieve the target of 60 dB SFDR for the Nyquist band.

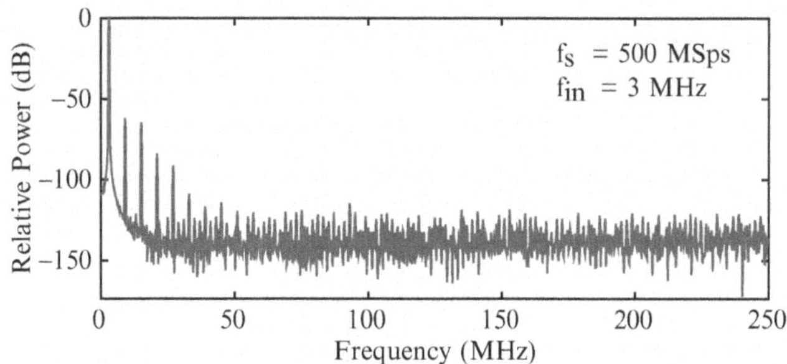

Figure 7.13. Output spectrum of the differential-pair based architecture.

Table 7.8 summarizes the simulated features of both T&H architectures, showing that both designs achieve the intended design goal. Even though the source follower based architecture achieves a better linearity, it was decided to implement the differential pair based design in a test chip as it is more suitable for the calibration methods that will be introduced in Chapters 8 and 9. Figure 7.16 shows the layout of the implemented T&H, measuring 90×90 μm.

Table 7.8. Features of the open-loop T&H circuits.

	Source follower	**Differential pair**
Sampling speed	500 MSps	
Static accuracy	68–77 dB	62 dB
Dynamic accuracy (1 GHz)	44–53 dB	52 dB
Power consumption	14.4 mW	

7. Experimental results

7.1 Measurement setup

When measuring the output of a T&H circuit, it is important to realize that the function of a T&H is to convert a continuous-time input signal into a sampled (or discrete-time) output signal. An illustration of this process is given in Fig. 7.17: during the *track phase*, the output of the T&H follows the input signal, which is a sinusoid in this example. Then, during the *hold phase*, the output of the T&H is held at the value which was sampled at the beginning of the hold phase. The series of held values compose the discrete-time signal generated by the T&H, as indicated by the *samples* in the figure. As the output

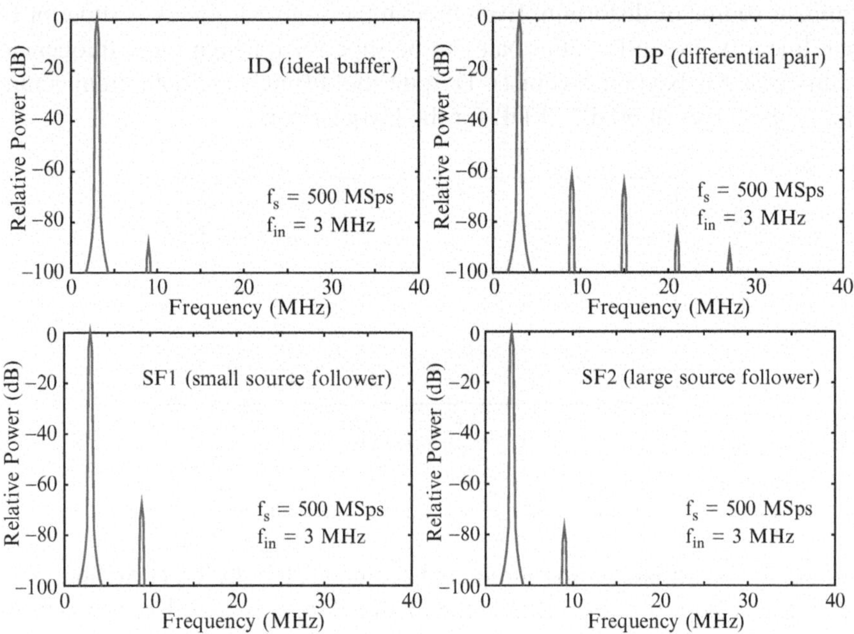

Figure 7.14. Partial output spectrums of the four architectures of Table 7.7.

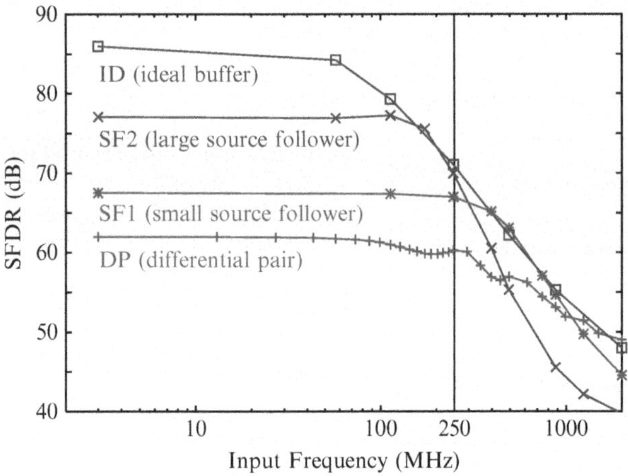

Figure 7.15. SFDR (linearity) as a function of the input frequency.

information is fully contained within the hold phase, the output during the track phase is irrelevant for the performance of the T&H. Therefore, when measuring the T&H, one should use the output during the hold phase only. This can be achieved by using an ADC to sample the output of the T&H in the hold phase

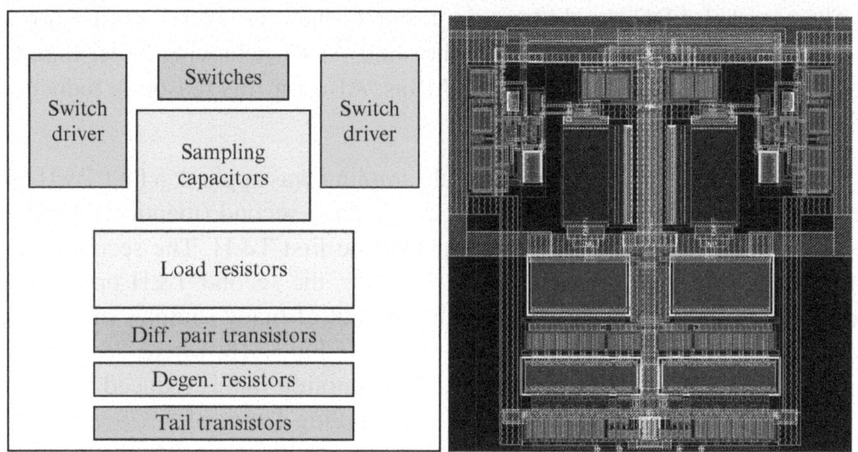

Figure 7.16. Floorplan and layout of the T&H. The shown area is 90×90 μm.

as indicated in Fig. 7.17. Then, the digital output data can be used to determine the performance of the T&H.

During the experiments, an off-chip ADC had to be used instead of an integrated one. This causes several problems for the measurements, namely:

- The T&H, while operating at 500 MSps, cannot drive the parasitic load associated with the package, PCB and external ADC. Because of that, it is not possible to transfer the output samples off-chip at the full sample rate of 500 MSps.

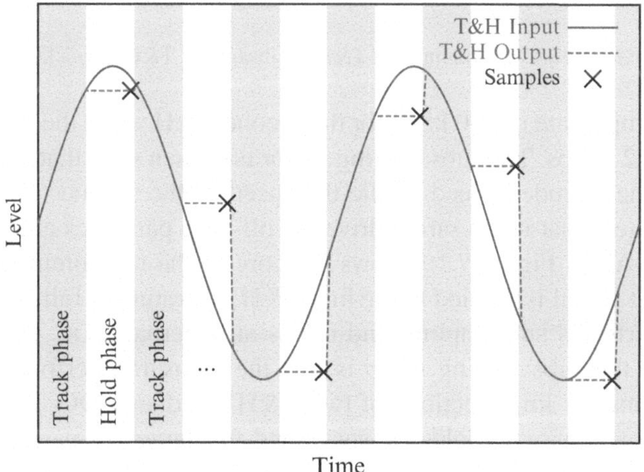

Figure 7.17. Ideal sketch of input and output signals of the T&H and sampling instants.

- The external ADC should be more accurate than the T&H in order not to limit the measurement result. Off-the-shelf ADCs achieving better than 60 dB SFDR do exist, but not at 500 MSps. Also for this reason, a reduction of the sample rate is necessary.

To work around these issues, on-chip subsampling was applied: a first T&H operates at the intended speed of 500 MSps. Then, a second (identical) T&H in the same chip is connected to the output of the first T&H. The second T&H samples the output of the first T&H. However, the second T&H operates at a lower sampling rate by resolving only one out of many samples of the first T&H. An illustration of this process is shown in Fig. 7.18, where T&H 2 subsamples T&H 1 by a factor of two (i.e.: the sampling rate is reduced by a factor of two). During the experiments, a subsampling factor of 3,328 was used,

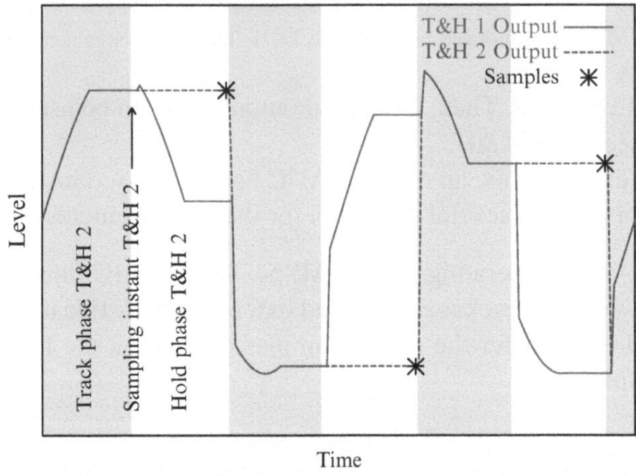

Figure 7.18. Ideal sketch of subsampling: T&H 2 subsamples T&H 1 by a factor of 2.

yielding a sample rate of 150 kSps for the second T&H while the first T&H operates at 499.2 MSps. The subsampling factor is chosen such that the first T&H operates at the intended speed, while the speed of the second one is reduced to such an extent that it can safely drive the off-chip parasitics and an external off-the-shelf ADC. Figure 7.19 shows the core of the measurement setup: an external input signal is applied to the first T&H, operating at full-speed. A second T&H performs subsampling and drives an external ADC. An important drawback of the subsampling setup is that the measured performance at the output contains the imperfections of two T&H's and an ADC. Therefore, the experimental results will yield a lower bound for the actual performance of a single T&H.

A custom setup was put together for the T&H measurements, as shown in Fig. 7.20. A computer running Matlab is used to control the automated

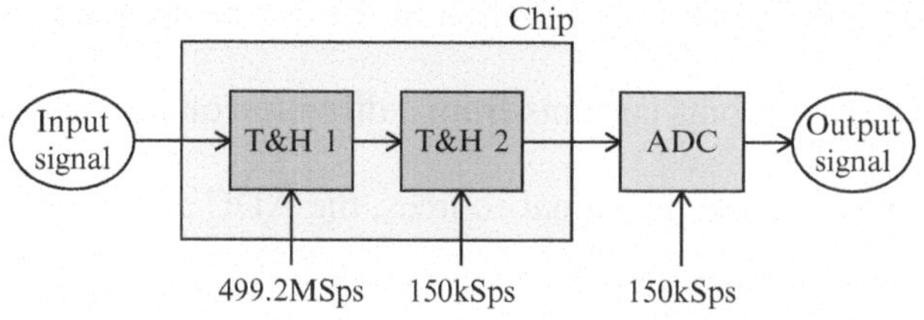

Figure 7.19. Subsampling method to facilitate off-chip AD conversion.

setup. The signal generator, clock generator and power supplies are controlled through a GPIB connection. The signal generator produces a sinusoidal signal which is applied to the T&H after low-pass filtering and single-ended-to-

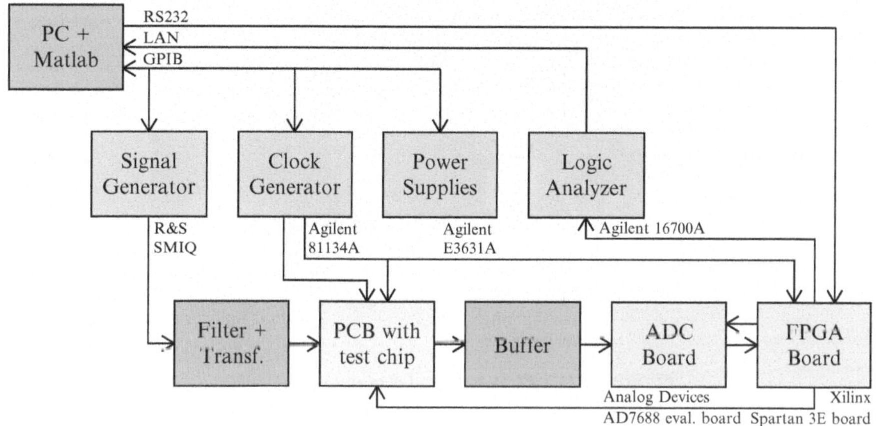

Figure 7.20. T&H measurement setup.

differential conversion. The low-pass filtering is used to improve the linearity of the input sinusoid as the linearity of the generator is less than 60 dB. The clock generator produces both the 499.2 MHz and the 150 kHz clock, used for the T&H circuits. The second T&H is connected to the external ADC through a buffer stage; the buffer is used to shift the common-mode voltage to a desirable level for the ADC. An FPGA board is used to control the ADC, to store the ADC output data to a logic analyzer and to control general settings in the test chip. The FPGA is also controlled from the computer, for example to initiate a measurement, to change a setting in the test chip or to start storing data to the logic analyzer.

Next to the custom measurement setup, a custom PCB is required to interface to the test chip. A photo of the board is shown in Fig. 7.21. The PCB implements several tasks:

- Local decoupling and low-pass filtering of the supplies and bias connections.

- Local (100 Ω) differential termination of the clock and input signals.
- Local generation of bias currents from a discrete voltage reference.
- Providing interfaces to the signal sources, the ADC and the FPGA.

Figure 7.21. PCB for the T&H measurements.

Because of the high speed of operation and the required accuracy levels, a proper implementation of the PCB is important. Key-aspects taken into account are:

- The input signal and clock signals are routed on top of the PCB, without any interruptions or vias between the SMA connectors and the test chip. The tracks are designed for 50 Ω matching, terminated as close as possible to the chip and as symmetrical as possible.
- The supplies and bias connections are routed on the bottom-side of the PCB. As such, they do not intervene with the signals on top. Moreover, the decoupling capacitors can be placed exactly below the chip, which is as close as possible.
- The ground plane is uninterrupted in the signal directions.
- A clamp is used to hold the test chip (100 pin TQFP) in place. Preventing the need for a socket, the wiring length can be minimized.

7.2 Measurement results

For the evaluation of the dynamic performance of the T&H, the setup introduced in Fig. 7.19 was used. Full-scale input sinusoids of various frequencies were applied to the T&H. Based on the measurements, the SFDR and the ENOB (using the SNDR) were derived. The results are shown in Figs. 7.22 and 7.23, respectively. The SFDR (Fig. 7.22) remains above 59 dB throughout the Nyquist range. As visualized in the figure, the SFDR is limited by either second, third or fifth order distortion, dependent on the input frequency. This

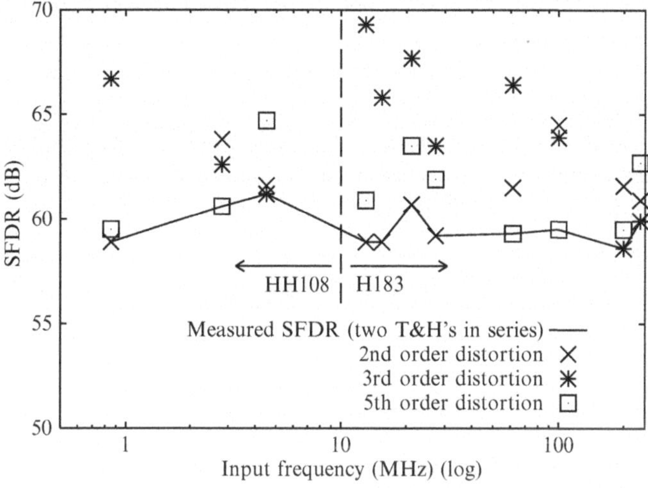

Figure 7.22. Measured SFDR of the T&H circuit, operating at 500 MSps.

dependency is not in agreement with the simulation results, where the linearity was invariably limited by the third order distortion component. The most likely cause of this effect is related to the network driving the input of the T&H, which is composed of the signal generator, low-pass filter, transformer, cables, PCB traces and termination network. In the presented implementation, the sampling switches of the T&H (see Fig. 7.2) are connected directly to the bondpads and the external network. At the sampling instant, the charge in the channel of the switch-transistor has to be drained; a part of this charge will flow into the sampling capacitor and a part will flow into the input network, which includes the external circuitry. The distribution of the charge depends on the impedances seen on both sides of the switch. Whenever the external impedance changes, it will affect the charge distribution which in turn can affect the linearity. Due to the limited bandwidth of the external components, its impedance will vary as a function of the applied frequency, which explains the obtained frequency-dependent SFDR. Most notably, two different transformers were used to convert the single-ended input signal to a differential one: a *Macom HH108* with a bandwidth of 200 kHz–30 MHz was used for the lower signal frequencies, and a *Macom H183* with a bandwidth of 30 MHz–3 GHz was used for the higher signal frequencies, as indicated in Fig. 7.22. From the measurements it appears that at the lower-end of the bandwidth of the transformers (0.85 MHz for the HH108 transformer and 13–27 MHz for the H183), the dominant distortion is of second order. This corresponds to the fact that for the lower-end of the bandwidth, the transformers exhibit an increased mismatch between the differential outputs, causing asymmetry in the input signal applied to the T&H. Because of that, even-order distortion dominates over odd-order distortion for these frequencies. In conclusion, it is most likely that the linearity is limited by the effect of the input network on the sampling structure and not by the differential-pair output buffer. Nonetheless, a 59 dB SFDR could be maintained, which is close to the intended 60 dB SFDR.

As shown in Fig. 7.23, the measured ENOB can be approximated by a simple first-order model that assumes 8.4 bit static accuracy and 3.8 ps clock jitter. This model can be expressed as:

$$ENOB = \left\{10\log_{10}\left((SNDR_{static}^{-1} + SNDR_{jitter}^{-1})^{-1}\right) - 1.76\right\}/6.02, \quad (7.23)$$

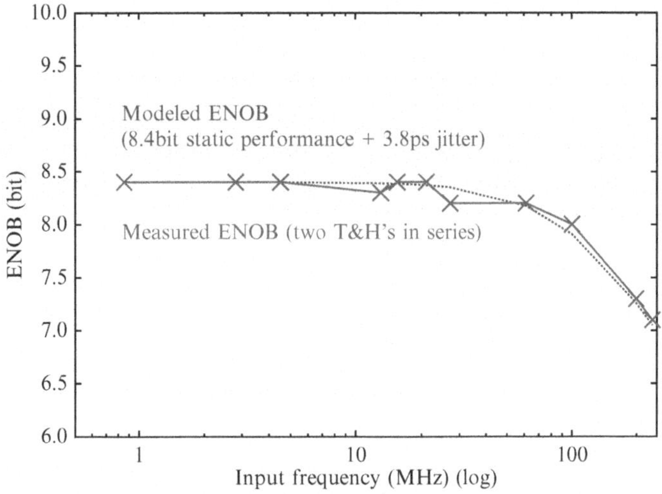

Figure 7.23. Measured ENOB of the T&H circuit, operating at 500 MSps.

where $SNDR_{static}$ models the static performance and $SNDR_{jitter}$ models the jitter performance. These terms can be expressed as a function of the static accuracy in bits (N_{static}), the time-skew (σ_t), and the input frequency (f_{in}):

$$\begin{cases} SNDR_{static} & = & 10^{(6.02N_{static}+1.76)/10} \\ SNDR_{jitter} & = & \left(\frac{1}{2\pi f_{in}\sigma_t}\right)^2 \end{cases} \tag{7.24}$$

To obtain the proposed model in Fig. 7.23, the parameters are set to $N_{static} = 8.4$ bit and $\sigma_t = 3.8$ ps. In the following, it will be explained why the experiments yield these figures.

First, the jitter performance will be investigated. The most likely cause of jitter is noise generated in the switch driver. Therefore, a transient noise analysis using the Mentor Eldo simulator was performed on the sampling core. During a transient noise analysis, one can observe the response to an applied input signal and the RMS noise voltage at each node, both as a function of time. Figure 7.24 shows the simulation results while sampling a 0.5 V, 247 MHz input sinusoid at 500 MSps. The top figure shows the signal output and

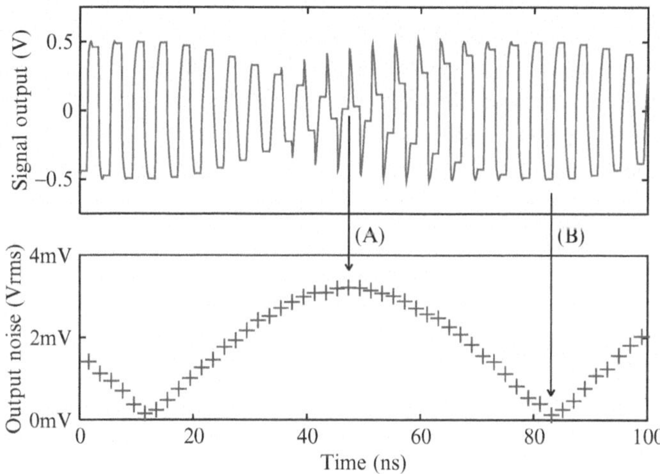

Figure 7.24. Simulated transient noise power (bottom) while sampling a 247 MHz sinusoidal input signal (top).

the bottom figure shows the RMS noise level for each held value. A maximum noise-level can be observed at sampling instant *(A)*. This is because at that moment, the input signal crosses zero; therefore, it reaches a maximum time-derivative, because of which it is most sensitive to jitter noise. A minimum noise-level can be observed at sampling instant *(B)*. At that moment, the input signal reaches its peak-level. As a result, the time-derivative will be close to zero and hence the jitter has less influence on the sampled signal. The resulting RMS noise figure is time-dependent and reaches a maximum of 3.1 mV_{rms}. Calculating the time average of the RMS noise levels results in an average noise-level of 2.2 mV_{rms}. As the input signal has a peak amplitude of 0.5 V, the achieved SNR at this frequency equals 44.1 dB. According to (7.24), this corresponds to $\sigma_t = 4$ ps, which is similar to the value of $\sigma_t = 3.8$ ps that was derived from the experimental results. Apart from jitter generated inside the chip, it is also possible that the measurement is limited by jitter coming from the signal generator or the clock generator. As the jitter specification of the clock generator is given to be 2 ps typical or 5ps maximum, this could also limit the measured result.

As a next step, the achieved static performance will be discussed. From the experiments, an ENOB of 8.4 bit could be observed, which corresponds to an SNDR of 52.3 dB. The SNDR considers both noise and distortion power. When splitting these terms from the measured data, the THD and SNR can be derived, yielding a THD of -56 dB and an SNR of 55 dB. The THD value is to be expected, as the measured SFDR (considering only the dominant distortion component) equals 59 dB. When adding all distortion components together, the resulting THD should be worse than, but relatively close to -59 dB. For the SNR, there are two contributions to be considered: the first one is kT/C noise from the sampling structure, the second one is the noise of the buffer stage. From Sections 4 and 5.4, it is known that both contributions are equal to 64 dB, resulting in a combined SNR of 61 dB. On top of that, two T&H circuits are used in series during the experiments, so the total input-referred SNR is expected to be 58 dB. The measured result of 55 dB is reasonably close to the expectation, but it cannot be excluded that other noise sources influence the measured performance.

A possible cause of the increased noise level is the presence of inductance in the signal path. A sketch of this situation is shown in 7.25: an external signal source connects to the sampling core through an input network. The input network is modeled by a resistor, an inductor and a capacitance. In reality, inductance is always present because of the bonding wires, package pins and PCB traces beyond the termination resistors. The capacitance models the parasitics related to the package and the bonding pad. It should be noted that the values of the components given in the figure are not based on a physical extraction of the implemented circuit; their only purpose is to show the principal behavior of the circuit. Because of the presence of the inductance, a ringing effect can be observed at the moment the switch samples the input signal. The characteristic frequency of the oscillation is dependent on the parasitics; as they are typically small, the oscillation will have a high frequency. For example, for L = 20 nH and C = 200 fF, the oscillation frequency is 2.5 GHz. Note that an inductance of 20 nH is relatively large, however, this value is selected to obtain a clear simulation result to illustrate the behavior of the circuit. The ringing effect introduces a *high-frequency* signal component at the sampling instant, even when the applied signal itself is of a low frequency. Now, the jitter of the

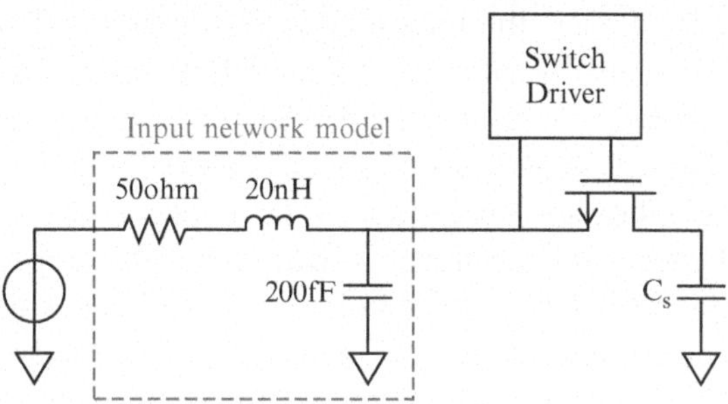

Figure 7.25. Sampling core with a model of the input network, including resistance, inductance and pad capacitance.

clock signal, which has normally negligible effect at low signal frequencies, becomes important again: it will modulate with the oscillation of the input network and add noise to the sampled data. Transient noise simulations were carried out on the circuit of Fig.7.25 to verify this principle. A low-frequency input signal of 4.3 MHz was used, which is normally insensitive to jitter noise. Figure 7.26 shows the applied signal and the simulated RMS noise. For comparison, the noise-plot is shown for a system with an inductance of 20 nH, and a system without inductance. Without inductance, an RMS noise-level of 0.18 mV_{rms} is achieved, which corresponds to the expected kT/C-noise level. However, when an inductance of 20 nH is present, there is a significant increase of the noise-level. The shown example reaches an average RMS noise-level of 0.7 mV_{rms}, which would produce an SNR of 54 dB.

A summary of the measured performance is included in Table 7.9. For comparison with prior art, the FoM-SFDR and FoM-ENOB are also calculated. As the T&H is designed for Nyquist operation, the ERBW should be equal to

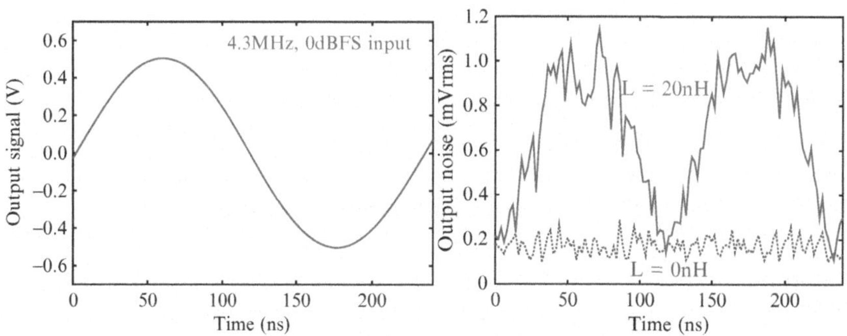

Figure 7.26. Simulated transient noise power (right) while sampling a 4.3 MHz sinusoidal input signal (left). Results are shown for a source inductance of 0 and 20 nH.

250 MHz. At this frequency, the measured ENOB equals 7.1 bit. As a 0.5 bit performance-loss is typically allowed at the ERBW, an ENOB of 7.6 bit can be claimed in combination with a 250 MHz ERBW. For the calculation of the FoM-ENOB, the latter ENOB is used as well. When comparing the measured result with the original design goal (Table 7.3), it appears that all requirements could be fulfilled except for the linearity which is 1 dB less compared to the target. However, the SFDR-based FoM of 37 fJ is still better than expected as the achieved power consumption of 13.5 mW is substantially better than the goal of 25 mW. Also, as explained previously, the most likely cause of the limited linearity is because of the measurement setup, and not because of the T&H itself.

Table 7.9. Measured performance of the T&H.

Power supply V_{DD}	1.8 V
Signal range $V_{in,pp}$	1.0 V
Power consumption	13.5 mW
$f_s, 2f_{in,max}$	500 MHz
SFDR (DC - Nyquist)	59 dB
Low-frequency ENOB	8.4 bit
ENOB at Nyquist	7.1 bit
ENOB for ERBW equal to Nyquist	7.6 bit
FoM-SFDR	37 fJ
FoM-ENOB	139 fJ

8. Conclusion

In this chapter, an open-loop Track-and-Hold circuit designed in a CMOS 0.18 μm technology was presented. Because of the open-loop architecture, a high speed of 500 MSps could be achieved. At the same time, by introducing a cross-coupled source degenerated differential pair, a high-linearity of 59 dB SFDR could be realized.

The experimentally verified performance in terms of the FoM-ENOB, the FoM-SFDR or the speed-linearity product is comparable to the current state-of-the-art. As such, the presented design provides a decent test-circuit for the calibration methods that will be presented in the following chapters.

The main performance limitation of the current implementation is due to jitter noise of the switch driver. This limits the ENOB at Nyquist to 7.1 bit, which is a loss of 1.3 bit compared to the static ENOB. By increasing the power consumption of the driver, the jitter noise could be reduced in order to improve the ENOB at Nyquist. As the power consumption of the driver (0.25 mW) is small compared to the overall consumption (13.5 mW), reducing the jitter would have little impact on the power budget. Therefore, it is expected that a further improvement of the FoM-ENOB is possible.

Chapter 8

T&H CALIBRATION

This chapter presents a method to enhance the accuracy of the open-loop T&H circuit, introduced in Chapter 7. The approach is able to measure offset, gain-error and non-linearity on-chip, and to correct for these imperfections by means of analog calibration. The calibration method will be discussed, and both simulation results and experimental results will be shown. Parts of this chapter have been published previously in [75, 76].

1. Introduction

In Chapter 7, a 500 MSps open-loop T&H was presented. The linearity of the proposed enhanced differential pair (Section 5.4) relies on the non-linearity of one differential pair compensating the non-linearity of another differential pair. Because of imperfections, the non-linearity compensation might not be exploited to the full extent. Even though a linearity of 59 dB could be verified experimentally, there are several issues which could result in a sub-optimal accuracy in practice:

- The transistor models might deviate from the actual transistor behavior.
- Because of process-spread, the nominal values of the components can deviate from the intended values.
- Random mismatch and systematic mismatch (e.g. die gradients or asymmetry in the layout) causes an additional variation of the values of the components.
- Environmental changes (temperature, bias current, etc.) modify the transfer characteristic.

As a result of these effects, the optimum solution according to simulations will not correspond exactly to the optimum solution in reality. In order to

P. Harpe et al., *Smart AD and DA Conversion*,
Analog Circuits and Signal Processing, DOI 10.1007/978-90-481-9042-3_8,

optimize the accuracy of the T&H given these unknown imperfections, a calibration method is proposed: iteratively, the calibration method measures the performance and optimizes it by tuning a set of parameters. Apart from optimizing the non-linearity, the method also optimizes the gain error and the offset.

In Section 2, the accuracy requirements for the T&H are discussed. Section 3 shows an overview of the calibration method, after which the details will be discussed in Sections 4 and 5. Simulation results and experimental results are shown in Sections 6 and 8, respectively, followed by conclusions in Section 9.

2. T&H accuracy

Apart from noise, the accuracy of a T&H is limited by non-linearity, gain error, and offset. The importance of these errors is dependent on the application of the T&H. For example, in time-interleaved ADCs gain and offset errors are important as they introduce unwanted spurious components. As another example, in a non time-interleaved ADC used in a low-IF receiver, offset and gain errors are not directly critical: the offset does not interfere with the signal, as there is no signal information at DC, and the gain error is typically compensated by an automatic gain-control loop.

As this concept study does not target a specific application, and it is desirable to provide a general method, all three imperfections will be taken into account. First of all, a set of relations is proposed to express the individual imperfections in terms of an effective number of bits. For the linearity, the effective number of bits ($ENOB_{SFDR}$) follows from the spurious free dynamic range ($SFDR$), as in Chapter 7:

$$ENOB_{SFDR} = (SFDR - 1.76)/6.02 \tag{8.1}$$

Assuming a nominal situation of zero offset and unity gain, the offset and gain error affect the transfer function as follows:

$$V_{out} = O_e + (1 + G_e)V_{in} = V_{in} + (O_e + G_e \cdot V_{in})\,, \tag{8.2}$$

where O_e equals the offset, G_e equals the gain-error and V_{in} and V_{out} represent the input and output voltage, respectively. The offset results in a constant deviation of the output, equal to O_e. On the other hand, the gain-error is multiplied by the input signal. For a given full-scale signal $\pm V_{fs}$, a maximum deviation of $G_e \cdot V_{fs}$ is realized. In this study, a maximum error equal to $\frac{1}{2}LSB$ of the $ENOB$ is tolerated.[1] As the LSB equals $2V_{fs}/2^{ENOB}$, the $ENOB_{offset}$ and $ENOB_{gain}$ can be derived:

$$ENOB_{offset} = \log_2\left(\frac{V_{fs}}{|O_e|}\right) \tag{8.3}$$

$$ENOB_{gain} = \log_2\left(\frac{1}{|G_e|}\right) \tag{8.4}$$

In the end, the overall *ENOB* is determined by the combined effect of all imperfections. Therefore, in line with the design goal from 2, the individual requirements are set to 10-bit, even though the overall ADC target is only 8-bit. Using relations (8.1), (8.3) and (8.4), the following set of constraints needs to be fulfilled in order to achieve 10-bit performance for a full-scale range of $V_{fs} = 0.5$ V:

$$\begin{cases} |O_e| & \leq & 0.5 \text{ mV} \\ |G_e| & \leq & 0.1\% \\ SFDR & \geq & 62 \text{ dB} \end{cases} \tag{8.5}$$

When the issues described in Section 1 (e.g. mismatch and process spread) are taken into account, these constraints can not be met for the T&H presented in Chapter 7. For example, consider the effect of random mismatch of the transistors composing the differential pair (Fig. 8.2) on the offset. Given an A_{vt} of 5 mV/μm^2, the standard deviation of the offset of the differential pair becomes: $\sigma = \sqrt{2}A_{vt}/\sqrt{WL}$. In order to meet the offset requirement (8.5) with a 3σ margin, the area of the input transistors should be 1,800 μm^2 (instead of the implemented 7 μm^2). First of all, this would have a serious impact on the total area of the T&H as these two devices alone would already occupy about half of the size of the current layout (Fig. 7.16). More important, for the intended speed of operation, it is not realistic to use such large devices as the parasitic C_{gs} equals 12 pF. This would give both speed and power issues in the preceding stage that has to drive the T&H. With the implemented small transistors, the parasitic load is reduced to 100 fF, which is smaller then the sampling capacitor. As such, the load for the previous stage is not limited by the T&H design, but by the sampling capacitor requirements. However, in case of these small devices, calibration of the offset (as well as the gain and the non-linearity) will be required. Therefore, a calibration method is presented to improve the intrinsic performance to the goal given in (8.5).

3. T&H calibration method

In line with the proposal in Chapter 6, the calibration method complies with the following properties:

- Foreground calibration method, performed at start-up.
- Digital processing of the information.
- Analog correction of the imperfections.

Figure 8.1 shows the calibration setup, implemented according to these properties. Being a foreground method, a dedicated test-signal is generated in the digital domain and applied to the T&H by means of a DAC. The analog output of the T&H has to be processed in the digital domain. In reality, the T&H is always followed by an ADC, so this ADC can be used as well to digitize the T&H's output during calibration. Then, the processing algorithm controls the analog parameters of the T&H based on the acquired data from the ADC.

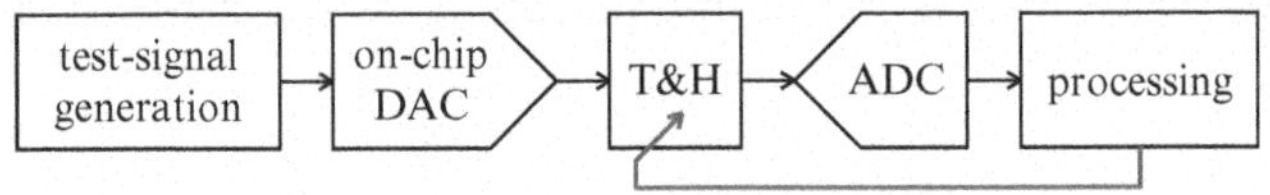

Figure 8.1. On-chip T&H calibration setup.

In order to have a successful implementation of the calibration method, suitable analog parameters are required to control gain, offset and non-linearity. Moreover, the imperfections should be measurable accurately, without relying on the accuracy of the test generator or the ADC. These two requirements will be discussed in the following two sections.

4. Analog correction parameters

For the optimization of the three relevant imperfections (offset, gain and SFDR), a minimum of three tunable analog parameters is required. However, the SFDR metric is determined by the combination of all distortion components. When multiple distortion components are critical for the obtained SFDR performance, a single parameter to optimize the SFDR might not be sufficient. Most notably, there is a difference between even-order distortion and odd-order distortion because of the differential nature of the T&H. This becomes clear from the example given in Fig. 8.2 which shows a simple differential pair:

- **Symmetrical changes** in the circuit affect the odd-order distortion, but have little influence on the second-order distortion as they maintain symmetry in the circuit. Examples of symmetrical changes are: changing the bias current I_B, or modifying the value of both resistors with the same amount.

- **Asymmetrical changes** in the circuit affect the even-order distortion as they create asymmetry in the circuit, but have less influence on the odd-order distortion. An example of an asymmetrical change is to increase the value of one resistor with a certain amount, and to decrease the value of the other resistor with the same amount.

To be able to optimize both even-order distortion and odd-order distortion, both symmetrical and asymmetrical changes have to be realizable. Because of that, a minimum of two parameters is required for SFDR tuning, which (together with offset and gain control) leads to a total of four tunable analog parameters.

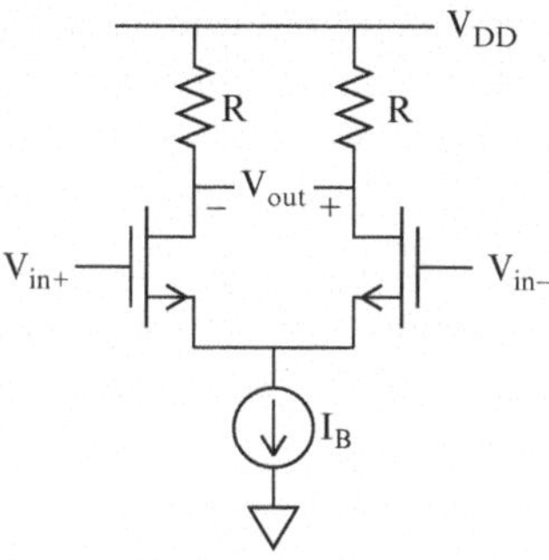

Figure 8.2. Differential pair with resistive load.

There are several alternatives how to implement the four programmable parameters in the T&H buffer (Fig. 8.3). It was decided to use the four tail current

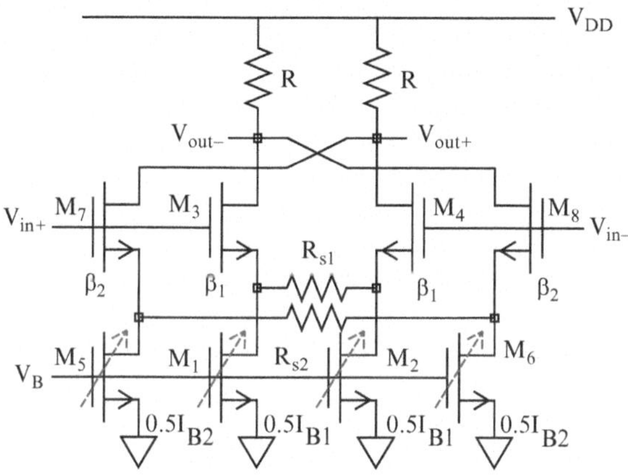

Figure 8.3. Enhanced differential pair with programmable tail current sources.

sources (implemented by transistors M_1, M_2, M_5 and M_6) as they have the following advantages:

- All imperfections that need to be optimized are sensitive to these tail currents. As such, these four parameters enable optimization of all the relevant imperfections.

- The current sources perform a static function in the buffer. Especially, the sources and gates of these devices are set to a constant level and only the drain potentials are signal dependent. Because of that, the programmable circuitry can be implemented in the static part of the buffer while the signal path remains free of additional switches or parasitics associated with the programmable components.

Each variable current source can be implemented as illustrated in Fig. 8.4: one large device (which is always turned on) provides a fixed current I_{fix}. A number n_{var} of binary-scaled devices ($n_{var} = 3$ in this example) is placed in parallel to the fixed source. These sources can be turned on or off individually by digital control signals, such that they realize a variable current between 0 and I_{var}. In this way, the programmable source can generate an output current in the range $[I_{fix}, I_{fix} + I_{var}]$ with $2^{n_{var}}$ discrete steps:

$$I_{total} = I_{fix} + \frac{p}{2^{n_{var}}} I_{var} , \tag{8.6}$$

where p is the digital control signal of the source, in the range $[0, 2^{n_{var}} - 1]$. By setting $I_{fix} + 0.5I_{var}$ to the original nominal value I_{nom} of the current source, a programmable range of $\pm 0.5 I_{var}$ around the nominal value can be realized.

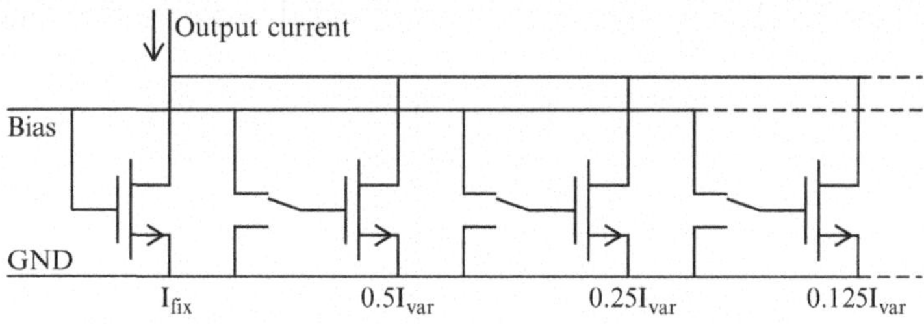

Figure 8.4. Implementation of the controllable current sources.

The two main design parameters of each programmable source are its range and its step size:

- The **range** determines the maximum imperfection that can be compensated. For example, the range of a current source corresponds to a certain range of gain-error that can be compensated. Taking into account e.g. random mismatch of components, possible process spread, the effect of gradients and layout asymmetry, an estimation can be made of the expected variation in gain. Based on that, the variable range of the current sources should be designed such that it covers the expected gain range. In this first prototype implementation, the range of the current sources was chosen relatively wide to ensure sufficient control.

- The **step size** determines with which accuracy the imperfections can be tuned, and determines the post-calibration accuracy. This limitation is inherent to the quantization effect of the variable current source, which results in an error up to 0.5 LSB or half the step size. As an example, according to the goal described in (8.5), the offset should be tuned to within 0.5 mV accuracy, which implies that the step size of the current sources should be chosen such that the offset can be controlled in steps of 1 mV at most. Similarly, the gain should be controllable in steps of 0.2% at most.

To determine the step size of the current sources of the main pair, consider the simplified single-ended view of a differential pair, as shown in Fig. 8.5. The

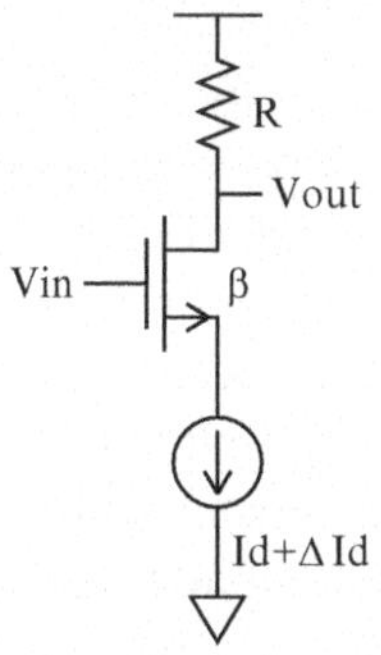

Figure 8.5. Simplified single-ended view of a differential pair.

effect of a small delta current ΔI_d on the offset and gain of this structure will be investigated. The gain of this stage, which is designed to equal unity, can be expressed as follows:

$$gain = \sqrt{2I_d\beta}R = 1 \tag{8.7}$$

The $\Delta gain$ that is introduced because of a small ΔI_d can be calculated by taking the derivative of (8.7):

$$\Delta gain = \frac{\partial gain}{\partial I_d}\Delta I_d = \frac{2\beta R}{2\sqrt{2I_d\beta}} = \frac{1}{2I_d} \cdot \Delta I_d \tag{8.8}$$

The offset of the structure is affected by the change in V_{eff}, as $V_{eff} = \sqrt{2I_d/\beta}$:

$$\Delta offset_{single} = \frac{\partial V_{eff}}{\partial I_d}\Delta I_d = \frac{1}{\sqrt{2I_d\beta}} \cdot \Delta I_d \tag{8.9}$$

For the differential pair, the introduced offset will be twice this value when the currents of both sides are adjusted at the same time (in different directions):

$$\Delta offset = \frac{2}{\sqrt{2I_d\beta}} \cdot \Delta I_d \tag{8.10}$$

As the nominal I_d is about 4 mA, and β is about 60 mA/V^2, relations (8.8) and (8.10) can be simplified and compared against the previously derived step-size requirements:

$$\begin{cases} \Delta gain & \approx & 125\Delta I_d & < & 0.2\% \\ \Delta offset & \approx & 90\Delta I_d & < & 1mV \end{cases} \tag{8.11}$$

From this, it follows that the step-size of the programmable current sources should be $\Delta I_d < 11$ μA. Allowing some margin, a step-size of 5 μA was selected for implementation which realizes gain steps of 0.06% and offset steps of 0.5 mV. As explained before, the range of control was chosen relatively wide in this first prototype. By using 8-bit programmable sources, 256 steps can be programmed, resulting in a control range of ±8% for the gain, and ±64 mV for the offset.

For the programmable sources in the cross-coupled pair, the design of the main pair was taken and scaled down corresponding to the difference in bias current between main and cross-coupled pair.

One issue when implementing the programmable current sources with transistors is the mismatch of these devices. Up to here, it is assumed that the 256 steps of each programmable current source are evenly distributed over its range. However, due to random mismatch of the transistors, there will be a random variation of the step size. The smaller the physical size of the transistors, the more pronounced the mismatch will be. Due to this mismatch, the step size can become larger than 1 LSB locally, which increases the quantization error for that specific value. As a result, the post-calibration performance can become worse than expected. Fortunately, the post-calibration performance will be degraded only when the mismatch causes an increased step size exactly at the required value of the programmable source.

Figure 8.6 shows Monte-Carlo simulation results on 10,000 buffers in which the offset is calibrated. The mismatches of the elements of the buffer are selected based on mismatch information from the technology. When no mismatch is present in the programmable current source, all 10,000 samples achieve a post-calibration offset less than 0.35 mV. Next, random mismatch is also added to the elements composing the programmable current sources. The amount of mismatch to be added is calculated by making an initial transistor-level implementation of the programmable current sources and

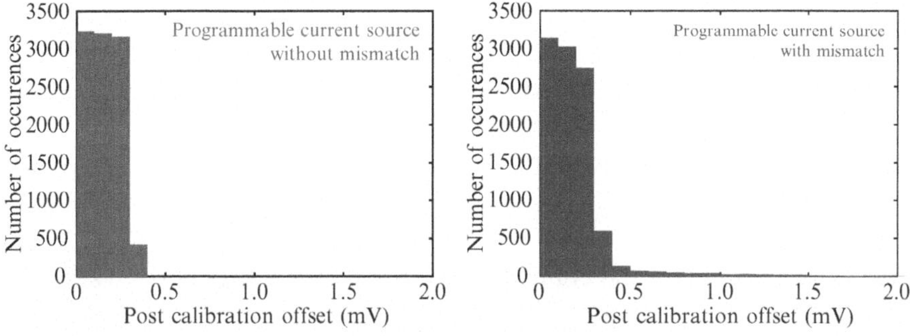

Figure 8.6. Post-calibration offset histogram without (left) and with (right) mismatch in the programmable source.

applying the mismatch model of the technology to that design. After adding this mismatch, the post-calibration performance of most of the samples remains similar to the original performance. However, for some samples the post-calibration performance is degraded due to the increased quantization error of the programmable source. Nonetheless, for 96% of all samples, the design goal (an error of less than 0.5 mV) can be achieved. As a yield of 96% is sufficient for a proof-of-concept, the initial transistor-level design of the programmable current sources was adopted for implementation. If necessary, the yield can be improved by increasing the area of the transistors composing the programmable sources, which reduces the mismatch. Alternatively, one could also consider to apply the redundancy method from Chapter 5 to the programmable current sources. By doing so, a smaller active area is feasible while it can be guaranteed that the quantization error remains below 0.5 LSB.

Based on the requirements for the step size, the range and the matching performance, the dimensions of the transistors composing the programmable current sources were selected: the $\frac{W}{L}$-ratio determines the current while the area WL determines the matching. The dimensions of the components are summarized in Table 8.1.

Table 8.1. Dimensions ($\frac{W}{L}$) of the programmable current sources.

	Main pair (M_1 and M_2)	**Cross-coupled pair** (M_5 and M_6)
Fixed:	$45.0\ \mu/180\ n$	$1.2\ \mu/180\ n$
Variable:		
Bit 7	$8\ \mu/250\ n$	$250\ n/250\ n$
Bit 6	$4\ \mu/250\ n$	$250\ n/500\ n$
Bit 5	$2\ \mu/250\ n$	$250\ n/1\ \mu$
Bit 4	$1\ \mu/250\ n$	$250\ n/2\ \mu$
Bit 3	$500\ n/250\ n$	$250\ n/4\ \mu$
Bit 2	$500\ n/500\ n$	$250\ n/8\ \mu$
Bit 1	$500\ n/1\ \mu$	$250\ n/16\ \mu$
Bit 0	$500\ n/2\ \mu$	$250\ n/32\ \mu$

To confirm the controllability of the imperfections by means of these four programmable current sources, Cadence simulations were carried out. Figure 8.7 shows how the transfer function of the buffer is affected by tuning the programmable currents in such a way as to maximize the offset or the gain-error. The nominal performance (with all programmable sources set to mid-scale) is also shown. From the figure, it can be observed that the programmable sources cover an offset range of ± 80 mV and a gain-error range of $\pm 8\%$. As the 256 steps of the 8-bit control are evenly distributed, this results in a step size of 0.6 mV for the offset and 0.06% for the gain. Knowing that the quantization error is half the step size at most, this corresponds to a post-calibration performance of ±0.3 mV for the offset and ±0.03% for the gain, which is within the target from (8.5). Also, the simulations correspond to the estimations calculated previously (offset step-size of 0.5 mV and gain step-size of 0.06%).

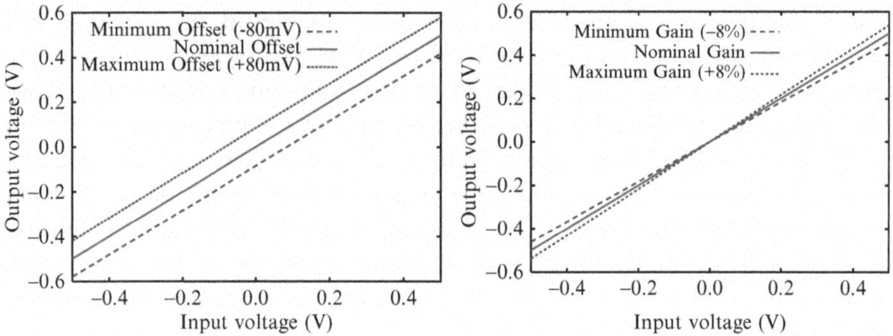

Figure 8.7. Offset and gain controllability.

Similarly, the even and odd-order distortion can be studied as a function of the programmable current. Figure 8.8 shows the distortion as a function of the input voltage, while tuning the parameters such as to maximize the even-order or odd-order distortion. The controllability becomes more clear when the distortion of the nominal design is subtracted from the total distortion. By doing so, only the distortion difference generated by the programmable sources remains. These results are shown in Fig. 8.9, which reveals that the programmable sources can control even-order and odd-order distortion independently.

In this section, the analog correction method was introduced for the compensation of offset, gain and non-linear imperfections. It was shown that sufficient

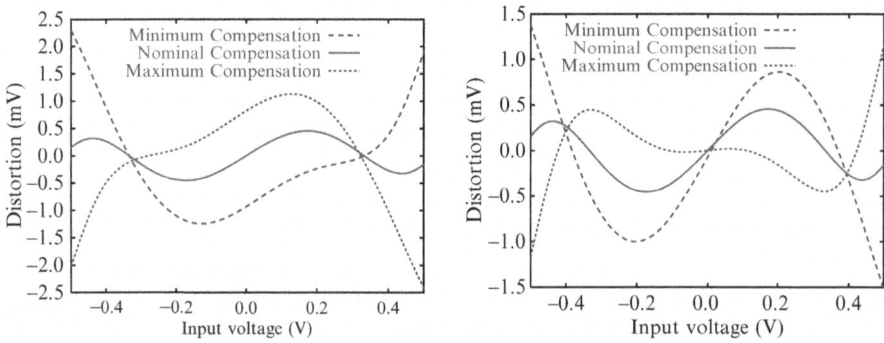

Figure 8.8. Even-order and odd-order distortion controllability.

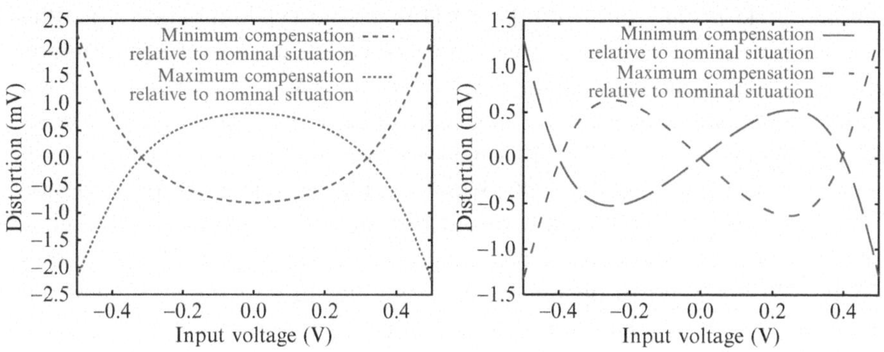

Figure 8.9. Relative even-order and odd-order distortion controllability.

accuracy and correction range can be implemented with simple programmable current sources. In the next section, the digital algorithm to measure the imperfections and to optimize the performance by means of the programmable sources will be discussed.

5. Digitally assisted analog correction

5.1 Self-measurement method

For the on-chip measurement of the imperfections of the T&H circuit, the setup from Fig. 8.1 is used. However, this setup contains three components with an unknown transfer function, namely the DAC, the T&H and the ADC. Each of these three components might produce offset, gain-error and distortion. For an on-chip calibration method, it is preferable that the accuracy of the self-measurement is not limited by the accuracy of the additional components used during the measurement; i.e.: the imperfections of the T&H should be measurable up to an accuracy which is beyond the accuracy of the DAC or the ADC. In order to do so, a small addition to the setup from Fig. 8.1 is made, as shown in Fig. 8.10: a switch is added at the input of the ADC. With this switch, the T&H can be either included or omitted from the measurement chain. An n_{dac}-bit DAC is used to generate a set of $2^{n_{dac}}$ static test signals. By means of the switch, either the analog input x or the output y of the T&H is selected and digitized by the ADC. In this way, each input code i (with $0 \leq i < 2^{n_{dac}}$) of the DAC results in two output codes of the ADC: one with the T&H inserted

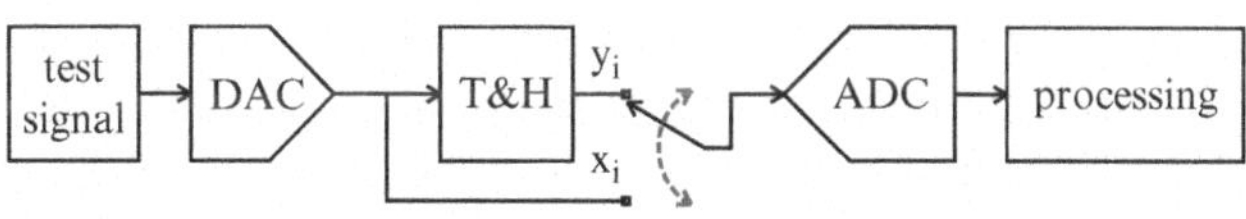

Figure 8.10. Self-measurement setup for the T&H circuit.

(yielding code y_i) and one with the T&H bypassed (yielding code x_i). When the T&H circuit is ideal, these codes must be equal: $x_i = y_i$. Each difference between x_i and y_i indicates an imperfection of the T&H. Note that these observations remain valid regardless of any static error in the DAC or ADC, like e.g. offset or distortion. Therefore, this measurement method is insensitive to the accuracy of the DAC and the ADC, which enables simple on-chip integration.

The performance of the switches used to connect the input of the ADC either to node x_i or y_i is important, as they are part of the signal path. However, these switches do not need to operate at the sampling rate of the ADC, but they remain in one setting during a measurement cycle. Because of that, they can easily achieve a performance (e.g. bandwidth, linearity) which is beyond the performance of the overall chain.

5.2 Optimization algorithm

The goal of the optimization algorithm is to minimize iteratively the differences between x_i and y_i by tuning the four programmable current sources. Instead of using a blind approach, the knowledge of the circuit and its behavior as a function of the parameters is used to reduce the complexity of the algorithm. Each iteration of the algorithm starts with a self-measurement cycle (Section 5.1). From these results, estimations of the different errors (offset, gain-error and distortion) are extracted. With these error estimations, the parameters controlling the four programmable current sources are updated. The four parameters for the four programmable current sources (Fig. 8.3) are indicated by p_1, p_2, p_3 and p_4 for transistors M_1, M_2, M_5 and M_6, respectively. After the parameter update, a new iteration is started until a stable solution is found. In the following, the error estimation will be discussed first, and then the parameter update.

As the measured codes x_i represent the input of the T&H and the codes y_i represent the output, the offset and gain-error can be estimated easily by means of a linear-fit in the least-squares sense using the data points (x_i, y_i). Then, only a residual signal d_i remains, which is the difference between the actual data points and the linear estimation:

$$d_i = y_i - \left(\hat{O}_e + (1 + \hat{G}_e)x_i\right), \tag{8.12}$$

where $\hat{O}_e$ is the offset estimation and $\hat{G}_e$ the gain-error estimation. The remaining difference d_i corresponds to the distortion of the T&H. As the required parameter update is different for even and odd order distortion, two separate quantities $\widehat{even}$ and $\widehat{odd}$ are used to quantify even and odd order distortion, respectively. For this goal, it is noted that the DAC is differential and produces both positive and negative levels of equal magnitude, such that the sum of these DAC levels equals zero:

$$\sum_{i=0}^{2^{n_{dac}}-1} x_i = 0 \tag{8.13}$$

In reality, the DAC might produce some offset at its output, which invalidates this assumption in theory. However, the effect of DAC offset can be neglected based on the following arguments: first, the main cause of offset in the DAC (implemented as described in Chapter 5), is mismatch in the output resistors. The specified resistor matching is better than 0.1%, which results in a 3σ output offset of less than 1 mV for the DAC, assuming a V_{fs} of 0.5 V. Assuming that a worst-case offset of 1mV would occur, this means that the DAC output range changes from $[-0.500$ V, 0.500 V] to $[-0.499$ V, 0.501 V]. Because of that, the non-linearity measurement and optimization will now be done for the $[-0.499$ V, 0.501 V]-range instead of the $[-0.500$ V, 0.500 V]-range. In practice, as the distortion is a smooth function across the entire signal range, this has no impact on the method.

Similar to the fact that (8.13) adds up to zero, all odd harmonic functions of x_i also add up to zero because of their symmetry. On the other hand, all even order harmonic functions of x_i add up to a non-zero value. The estimations $\widehat{even}$ and $\widehat{odd}$ are defined such that even order distortion contributes only to $\widehat{even}$, and odd order distortion only to $\widehat{odd}$:

$$\widehat{odd} = \sum_{i=0}^{2^{n_{dac}}-1} d_i x_i^3 \tag{8.14}$$

$$\widehat{even} = \sum_{i=0}^{2^{n_{dac}}-1} d_i x_i^2 \tag{8.15}$$

To understand the functionality of these estimations, the following situation is considered: suppose that the residual signal d_i contains an odd-order distortion component $C_a x_i^a$ (with a an odd number and C_a a constant). In (8.14) this term is multiplied by x_i^3, yielding $C_a x_i^{(a+3)}$, where $(a+3)$ is even. Therefore, the summation adds up to a non-zero value and contributes to $\widehat{odd}$. On the other hand, the same component $C_a x_i^a$ in (8.15) will be multiplied by x_i^2, yielding an odd-order term $C_a x_i^{(a+2)}$, which will add up to zero and therefore does not contribute to $\widehat{even}$. In a similar way, even order distortion components in d_i will contribute to $\widehat{even}$ only. In summary, with relations (8.14) and (8.15), it is possible to split the even-order and odd-order distortion components, which is necessary for the optimization algorithm that will be discussed next.

The final step of the optimization algorithm is to translate the extracted error estimations to updates of the parameters controlling the variable current sources. To reduce the complexity of this multidimensional problem (there are both four input signals ($\hat{O}_e$, $\hat{G}_e$, $\hat{even}$ and $\hat{odd}$) and four output signals (p_1 up to p_4)), available knowledge of the circuit is exploited. First of all, it is known that the T&H circuit is composed of two differential pairs (Fig. 8.3): one main pair (controlled by p_1 and p_2), which is responsible for the basic functionality of the buffer, and a much smaller cross coupled pair (controlled by p_3 and p_4), which has the task to compensate the distortion of the main pair. Consistent with this difference in functionality, the offset and gain errors ($\hat{O}_e$ and $\hat{G}_e$) are used only to update parameters p_1 and p_2, while the distortion estimations ($\hat{even}$ and $\hat{odd}$) are used only to update p_3 and p_4. In other words, the basic errors of gain and offset control the main differential pair and the non-linearity errors control the cross-coupled pair. With this procedure, the original four-dimensional problem is reduced to two two-dimensional problems. However, these two two-dimensional problems are not independent from each other. Because of that, the final solution found by the calibration method could be sub-optimal. Nonetheless, as shown later by simulation results, this simplified approach is able to find an optimum with sufficient performance.

Next to knowledge about the functionality of the circuit, knowledge about the relations between the parameters and the errors was taken into account. Based on circuit simulations, it can be concluded that the error estimations ($\hat{O}_e$, $\hat{G}_e$, $\hat{even}$ and $\hat{odd}$) are monotonous functions of the parameters p_1 up to p_4. In practice, as the mismatches are relatively small, these functions can be approximated by linear functions. Furthermore, the optimum solution is the solution where all error estimations are equal to zero. This means that in all cases, the sign of the error determines in which direction (positive or negative) the parameters should be updated. Under the assumption that the functions are linear, the magnitude of the update is automatically proportional to the error itself. Overall, this leads to the following update algorithm, where $p_x[k+1]$ is the new parameter, $p_x[k]$ the old parameter and $\Delta_x[k]$ the parameter update:

$$p_x[k+1] = p_x[k] + \Delta_x[k] \text{ , with:} \tag{8.16}$$

$$\begin{cases} \Delta_1[k] &= -\, c_1 \cdot \hat{G}_e[k] & -\, c_2 \cdot \hat{O}_e[k] \\ \Delta_2[k] &= -\, c_1 \cdot \hat{G}_e[k] & +\, c_2 \cdot \hat{O}_e[k] \\ \Delta_3[k] &= +\, c_3 \cdot \hat{odd}[k] & -\, c_4 \cdot \hat{even}[k] \\ \Delta_4[k] &= +\, c_3 \cdot \hat{odd}[k] & +\, c_4 \cdot \hat{even}[k] \end{cases} \tag{8.17}$$

One can see that in these equations, the gain and offset errors control p_1 and p_2 while the distortion errors control p_3 and p_4. The proportionality constants c_1 up to c_4 are chosen such that a fast and stable settling of the parameters can be achieved.

6. Simulation results

In this section, simulation results of the calibration method are presented. To verify the functionality of the calibration algorithm, Monte Carlo simulations were done on a behavioral model of the T&H. Next to that, simulations on transistor-level were performed to verify the achievable performance more accurately.

6.1 Behavioral-level simulations

In order to be able to run Monte Carlo simulations, a behavioral model of the open-loop amplifier from Fig. 8.3 was developed in Matlab. Most importantly, the four controllable current sources were modeled as (8.6). The four transistors composing the two differential pairs were modeled by the relation $I_d = \frac{1}{2}\beta(V_{gs} - V_{th})^2$. The required measurement DAC was modeled as a 6-bit binary DAC, including a mismatch of $\sigma = 5\%$ of the unit elements. Mismatch was added to all components of the amplifier, according to specifications of the technology. A Monte Carlo analysis was performed on 100 circuits. Each circuit was optimized by means of the presented self-measurement and correction algorithm. The achieved performance was validated both before and after optimization by applying an ideal input sinusoid and deriving the gain-error, offset and THD from the output data. Figures 8.11, 8.12 and 8.13 show the results before and after optimization. The nominal performance (achieved for a circuit without mismatch after parameter optimization) is also shown.

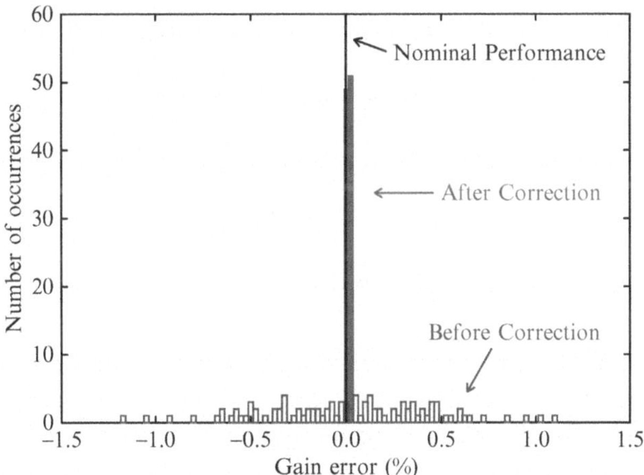

Figure 8.11. Gain error before and after correction.

Using (8.3) and (8.4), it can be concluded that before correction, the offset and gain errors limit the performance to 6 or 7-bit accuracy, while after correction, a performance of more than 10-bit is achieved, in line with the design

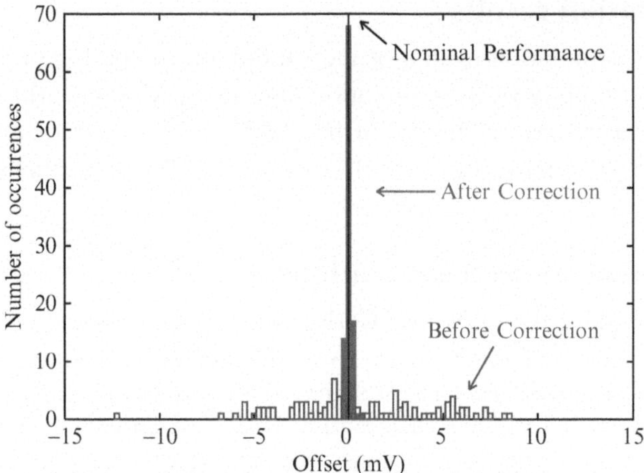

Figure 8.12. Offset before and after correction.

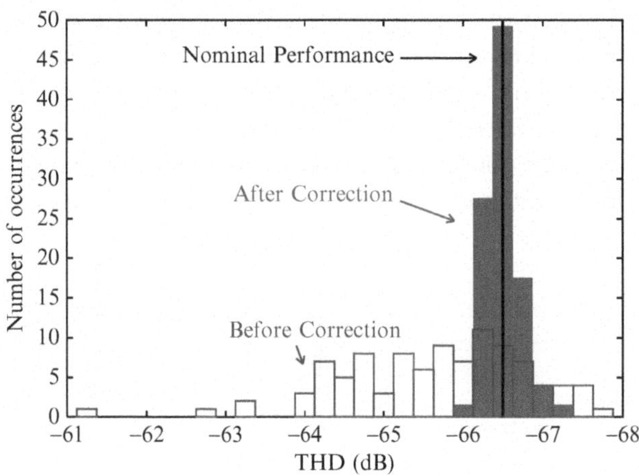

Figure 8.13. THD before and after correction.

goal. Even the outliers can be corrected fully by means of the proposed method. The THD figure shows that despite the large range of mismatch, the THD is compensated to at least −66 dB. In few cases, the post-calibration THD is slightly worse compared to the pre-calibration THD. This is because gain and offset are tuned at the same time. When gain and offset are being changed, this will also affect the non-linearity. On top of that, the 67 dB THD level is already beyond the intended performance of 62 dB.

6.2 Transistor-level simulations

Simulations were performed on a full transistor-level implementation of the T&H circuit as well. Because of computational limitations, only four simulations were carried out. First of all, the optimization algorithm was used on the nominal T&H (without mismatches). Next to that, three simulations were performed with different combinations of mismatch. In each of these cases, mismatch was added to each of the components of the open-loop buffer, being: the four resistors, the four transistors composing the differential pairs and the transistors implementing the four variable current sources. Based on technology information, the σ of each of these components was derived. Extreme mismatch cases were simulated by adding a mismatch of either -3σ or $+3\sigma$ to each component, and choosing only the sign of the mismatch randomly for each component. This approach was repeated three times, resulting in the three mismatch simulations. In all cases, a stable parameter solution was found within 32 iterations of the algorithm. Table 8.2 summarizes the results, showing the gain-error, offset and THD both before and after correction. For convenience, the errors are also expressed in equivalent accuracy according to equations (8.1), (8.3) and (8.4). It can be seen that gain and offset errors limit the initial performance to 5 or 6-bit accuracy, but after optimization an accuracy of more than 11-bit is achieved. The linearity in terms of THD improves with around 6 dB or 1 bit to -63 dB. Overall, the T&H achieves the 10-bit performance goal.

Table 8.2. Extreme-case transistor-level simulation results.

	Gain error		**Offset**		**THD**	
	%	bit	mV	bit	dB	bit
Nominal	0.00	∞	0.00	∞	-64.0	10.3
Before calibration						
Mismatch 1	1.32	6.2	8.90	5.8	-58.6	9.4
Mismatch 2	1.65	5.9	12.53	5.3	-57.9	9.3
Mismatch 3	0.29	8.4	-16.86	4.9	-54.8	8.8
After calibration						
Mismatch 1	0.04	11.2	-0.09	12.5	-62.4	10.1
Mismatch 2	0.04	11.4	0.07	12.8	-63.8	10.3
Mismatch 3	0.02	12.4	-0.12	12.0	-63.1	10.2

7. Implementation of the calibration method and layout

For the experimental verification of the calibration method, several components could be reused:

- The T&H from Chapter 7.
- The DAC from Chapter 5, used as test-signal generator.

On top of that, several additions were made to the design to facilitate the calibration method:

- The programmable current sources are added to the T&H as well as local digital registers to store the parameter values.
- Bypass switches are added to the T&H, as in Fig. 8.10.
- Switches are added to the input of the T&H to either use the DAC as input signal (during calibration), or the external input (during normal operation).

Figure 8.14 shows a photograph of the implemented test chip in a CMOS 0.18 μm process. The chip contains two programmable T&H's and a DAC. The T&H's (85 × 85 μm each) and the DAC (330 × 115 μm) were optimized for area, but the programmable current sources (275 × 70 μm, including digital control logic) were not optimized. Out of the 275 × 70 μm, 20 × 70 μm is used by the analog current sources, while the remaining part is used by the (manually designed) flip-flops. With area optimization, it should be feasible to use about 10 × 10 μm per flip-flop, or 80 × 40 μm for the total of four 8-bit flip-flops. Then, the size of the programmable part reduces from 275 × 70 μm to 80 × 60 μm. Also, the implemented DAC has 16-bit resolution, while the calibration method requires only 6-bit. A redesigned DAC for 6-bit resolution would approximately measure 120 × 70 μm.

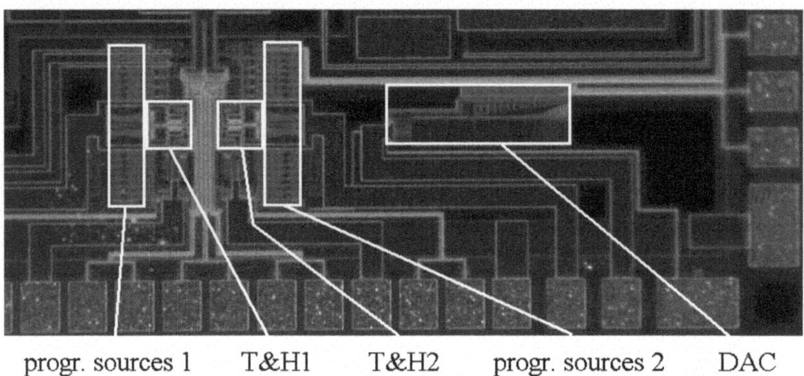

Figure 8.14. Photograph of the T&H structure and DAC. The shown area is approximately 1,450 × 600 μm.

8. Experimental results

8.1 Measurement setup

For the experimental verification of the calibration method, a dedicated measurement setup was created. For the self-measurement phase of the calibration method, the setup in Fig. 8.15 was implemented: a 6-bit digital ramp is generated inside an FPGA and applied to the DAC that is available inside the test-chip. Even though the implemented DAC (Chapter 5) has a resolution of 16-bit, only 6 bits are used during these experiments. Then, the analog test signal is applied to one of the two T&H's that is available inside each test-chip. The output is sampled by an off-chip ADC and processed externally to run the calibration algorithm. From the 24-bit ADC output, only the 10 MSBs are used to emulate a realistic ADC performance. Because no sub-sampling is applied during these measurements and a low-speed ADC is used, the sampling rate is fixed at 3kSps.

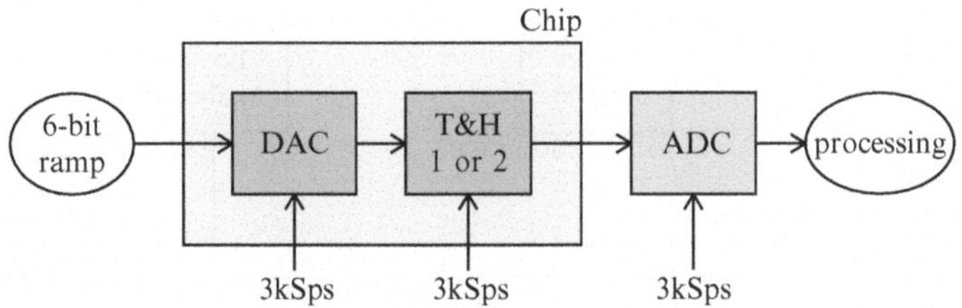

Figure 8.15. Setup for self-measurement of T&H imperfections.

For the verification of the achieved performance of the T&H, both before and after calibration, the setup in Fig. 8.16 is used: an external reference-ramp is generated which has an equivalent accuracy of at least 12-bit. This signal is fed to one of the two T&H circuits and sampled by the off-chip ADC. From the ADC output, the transfer function of the T&H can be obtained and the imperfections (offset, gain-error and distortion) can be determined. It should be noted that only the static performance was verified. An evaluation of the dynamic performance could not be performed due to the fact that the dynamic performance is limited by the measurement setup and not by the T&H (as explained in Section 7). Because of that, calibration of the T&H has no impact on the measured dynamic performance.

Figure 8.17 shows the equipment used for both the self-measurement phase and the verification phase. A custom mixed-signal board was designed around the Analog Devices AD1980 Codec. This chip includes both AD and DA

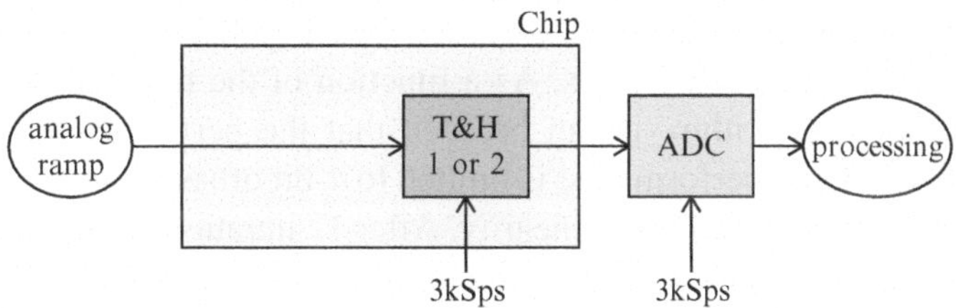

Figure 8.16. Setup for T&H performance verification.

converters: one ADC is used to process the output of the T&H while one DAC is used to create the ramp signal during verification mode. At the same time, the board provides the clock signals for the T&H and the FPGA. The AD/DA board is controlled by a Spartan 3 board, which also transmits the output data to the logic analyzer for further processing on the PC. A second FPGA board (Spartan 3E) is used to control the custom settings inside the test-chip.

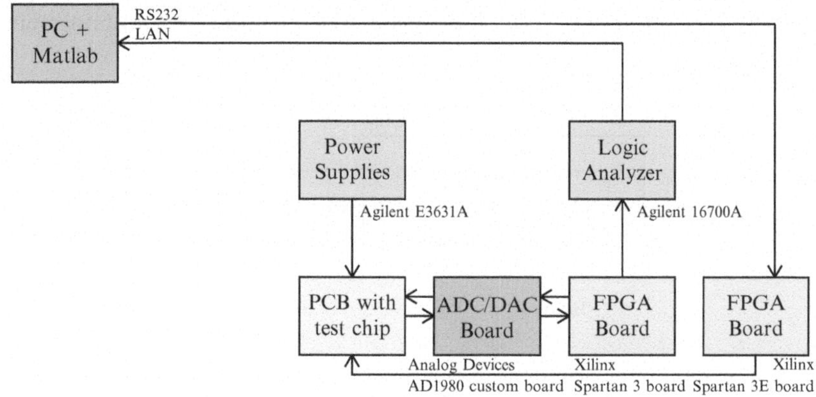

Figure 8.17. T&H measurement setup.

8.2 Measurement results

For the determination of the static performance, the transfer curve of the T&H circuit is measured using a low-frequency ramp signal. From the measured curve, gain, offset and distortion can be determined. The distortion is expressed in terms of SFDR, by estimating the frequency spectrum of the T&H in software, based on the measured DC characteristic. As each chip contains two T&H's, the results for both circuits will be given.

The presented analog calibration method was applied to the test chip. As the method updates the programmable current sources iteratively, the performance of the T&H can be evaluated after each cycle of the algorithm. The measured results for offset, gain-error and SFDR are given in Figures 8.18, 8.19 and 8.20, respectively. While offset and SFDR can be measured in an absolute sense, the gain-error is a relative error. Because of that, Fig. 8.19 shows the relative gain-error between the two T&H's. As a function of the number of iterations of the calibration algorithm, it can be seen that the performance improves. Before calibration, the performance is limited to 4-bit offset performance, 5-bit gain-error performance and 9-bit linearity. After 12 iterations of the calibration algorithm, the performance is improved to 10-bit or 11-bit, which corresponds to the simulation results and matches the 10-bit accuracy goal.

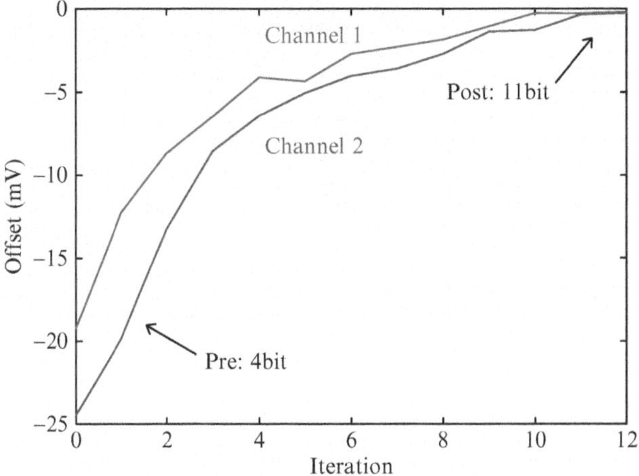

Figure 8.18. Measured offset of the two T&H's.

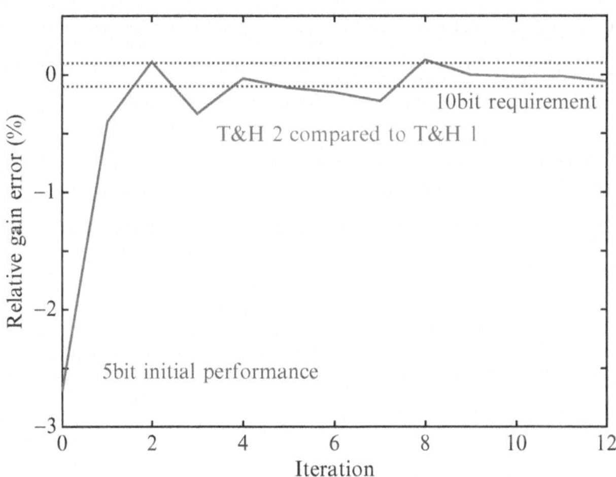

Figure 8.19. Measured relative gain-error of the two T&H's.

The development of the linearity as a function of the iteration step (Fig. 8.20) is not monotonous. There are several reasons that can cause the fluctuations during the linearity optimization:

- At the same time, the circuit is optimized in multiple dimensions: gain, offset, even-order distortion and odd-order distortion. When the gain and offset are being tuned, fluctuations in the linearity can be expected.
- The controllable current sources to tune gain, offset and linearity are based on small transistors. Thus, the transfer function realized by these

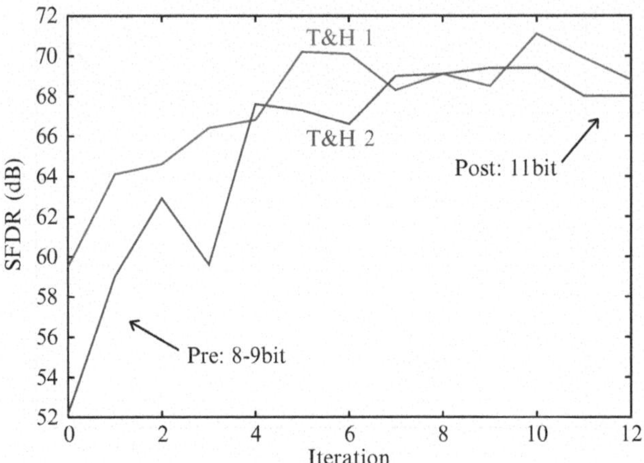

Figure 8.20. Measured linearity of the two T&H's.

controllable current sources might be non-monotonous. Even though the feedback algorithm works properly even with this non-monotonicity, it can cause fluctuations during the performance optimization.

- As the calibration method is performed off-line, the total calibration procedure takes several hours. In that period, environmental changes (e.g. temperature, supply, biasing) can cause fluctuations in the performance.

9. Conclusion

In this chapter, a method for the on-chip measurement and correction of offset, gain-error and distortion of the open-loop T&H circuit was presented. The method is suitable for on-chip implementation, as it does not require an accurate reference source or an accurate measurement device. The actual correction is performed in the analog domain, such that no additional processing power is consumed at runtime. Extensive simulations confirm a performance improvement of 5bit with respect to gain and offset errors, and a linearity improvement of 6dB. Experimental results confirm that a 10-bit post-calibration performance can be achieved. As a result, next to high-speed and low-power operation, also high-accuracy can be achieved by the open-loop T&H circuit without increasing the power consumption.

Notes

1. Note that this is a rather arbitrary choice to demonstrate the concept; in practice, the requirement should be set dependent on the target application.

Chapter 9

T&H CALIBRATION FOR TIME-INTERLEAVED ADCS

This chapter presents a method to enhance the performance of a time-interleaved ADC, using the open-loop T&H circuit introduced in Chapter 7. The approach is able to measure the mismatches between the various channels of a time-interleaved system on-chip, and to correct for these imperfections by means of analog calibration. The calibration method will be discussed, and both simulation results and experimental results will be shown. Parts of this chapter have been published previously in [77, 78].

1. Introduction

Time-interleaving multiple analog-to-digital converters (ADCs) [59] is a widely used approach to accommodate the demand for higher sampling rates combined with high accuracy and low power consumption. For example, in Fig. 9.1 p parallel ADCs, each with a separate track-and-hold (T&H) circuit, are combined to compose a p times faster ADC: each T&H samples the same input signal V_{in} with the same sample rate $\frac{f_s}{p}$, but the samples are taken at different phases of the clock, such that after digital recombination, the overall system behaves as a single ADC operating at f_s. In this work, the combination of a single T&H and a single ADC will be called a *channel*; i.e. the complete ADC is then composed of p channels. Apart from the architecture in Fig. 9.1, where each channel contains its own T&H, there are also alternative solutions like using one dedicated T&H in front of all channels. However, this work focusses on the architecture of Fig. 9.1, which is a commonly used solution.

The accuracy of a time-interleaved ADC is limited by two properties, namely: the accuracy of the individual channels that compose the time-interleaved ADC, and the matching accuracy between the channels. When the open-loop T&H as presented in Chapter 7 is to be used in a time-interleaved converter, both the accuracy of each individual T&H and the matching between

P. Harpe et al., *Smart AD and DA Conversion*,
Analog Circuits and Signal Processing, DOI 10.1007/978-90-481-9042-3_9,

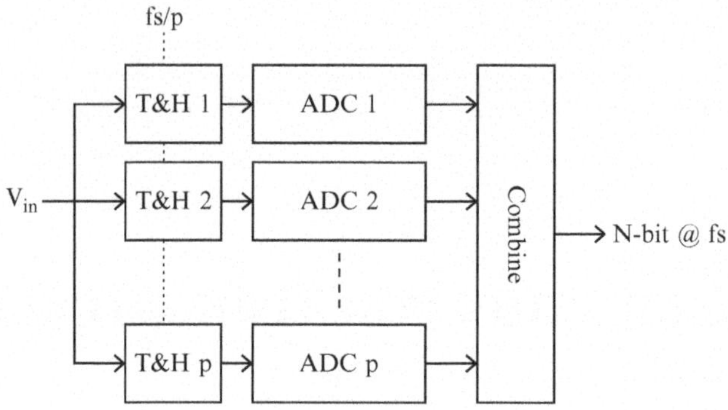

Figure 9.1. N-bit p-channel time-interleaved ADC.

several T&H's has to be taken into account. The first requirement can be fulfilled by the calibration method presented in Chapter 8. Therefore, this chapter focusses on the calibration of the matching errors between the T&H channels.[1] In practice, a combination of both calibration methods could be implemented to correct for both the errors of the individual channels and the errors between the different channels. However, this combination is beyond the scope of this work.

There are many effects that result in matching errors between the ideally identical channels, for example:

- Random mismatch of components (transistors and capacitors) in the channels and the clock circuitry.
- Systematic mismatch of components due to gradients on the die, affecting the channels differently.
- Timing mismatches due to differences in wiring and capacitance in the common clock circuitry and input network.
- Common-mode and power supply gradients due to differences in DC paths.

As all these problems are dependent on the architecture of the channels, the transistor-level design, the technology properties and the actual layout, it is difficult to derive an exact model of the mismatch errors. Instead, it is common practice to use an abstract model to represent the mismatch errors. The most widely used and accepted model considers three mismatch errors: offset, gain error and time-skew [60]. On system-level, this error-model can be used to derive mismatch requirements based on the final accuracy target. Then, during the design and implementation of the ADC, the actual errors can be translated to this simplified model and verified accordingly.

Normally, the mismatch errors (offset, gain error and time-skew) directly limit the performance of the overall ADC, and should be small enough to achieve the final speed/accuracy target. However, especially for high sampling rates and a large amount of channels p, it is difficult to achieve sufficient matching by design alone. Therefore, solutions were developed that can improve the channel-matching by measuring and correcting the actual errors on-chip (e.g. [79–83]). Several distinct properties can be found within the currently available techniques, but most of them share one or more of the following disadvantages or limitations:

- The method works only for a subset of the three types of mismatch (offsets, gain errors and time-skew errors).
- The method puts constraints on the input signal (background techniques), or requires input signals with specific accuracy requirements (foreground techniques).
- The method works only for a 2-channel ADC, or the complexity increases strongly (i.e. faster than a linear increase) as a function of the number of channels.
- The complexity (and hence the power consumption) of the correction method is such that it becomes unattractive for a power-efficient implementation.
- The method is based on a stochastic process and therefore requires a large amount of observations or iterations to achieve a certain level of accuracy.

Because of these drawbacks, a new method is proposed here that has the following properties:

- It measures and corrects for offsets, gain errors and time-skew errors.
- The foreground method uses a deterministic test-signal, of which the accuracy and dynamic performance are unimportant, as long as the signal is periodic. The low constraints enable a simple on-chip implementation.
- Because of the deterministic nature, the algorithm converges fast.
- The implementation is such that it can be applied to any number p of parallel channels, while the complexity grows only linear with p.
- The actual correction is performed by means of analog calibration to minimize the additional power consumption.

Section 2 reviews the requirements on channel matching for time-interleaved T&H's. In Section 3, the time-interleaved calibration system is presented, including the test-signal generator and a model of the error mechanisms. Sections 4 and 5 discuss the error detection and error correction schemes, respectively. Simulation results are given in Section 6. The hardware implementation is reviewed in Section 7 and experimental results are shown in Section 8. Finally, conclusions are drawn in Section 9.

2. Channel matching in time-interleaved T&H's

The effect of channel mismatch in time-interleaved ADCs has been described extensively in literature, a.o. [60]. For the specific design in this work (a 2-channel time-interleaved system, operating at 1 GSps, $1V_{pp}$ signal range), the effect of offset, gain-error and time-skew was investigated. The two channels were modeled according to (9.1), assuming that the input signal is a sinusoid with amplitude A and frequency f. Moreover, O_e, G_e and Δ model the offset, gain-error and time-skew, respectively.

$$\begin{cases} y_1 &= A \cdot \sin(2\pi f t) \\ y_2 &= O_e + (1 + G_e) \cdot A \cdot \sin(2\pi f (t + \Delta)) \end{cases} \tag{9.1}$$

As a function of the applied mismatches (either offset, gain-error or time-skew), the ENOB of the converter can be determined as shown in Fig. 9.2, using $A = 0.5$ V and $f = 487$ MHz. By using a full-scale input signal and a frequency close to Nyquist, these results correspond to a worst-case scenario. For a 10-bit accuracy goal, this results in the following set of requirements on channel matching, when the errors are considered individually:

$$\begin{cases} |O_e| &\leq 0.5\text{mV} \\ |G_e| &\leq 0.2\% \\ |\Delta| &\geq 0.5\text{ps} \end{cases} \tag{9.2}$$

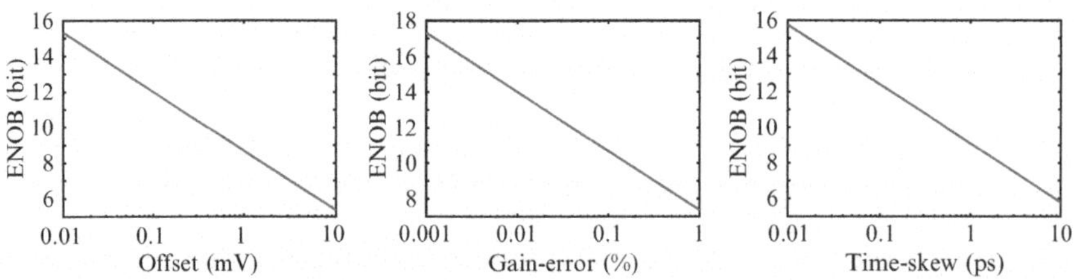

Figure 9.2. ENOB limitation due to mismatch (offset, gain-error and time-skew).

3. Channel mismatch calibration

3.1 System overview

Figure 9.3 shows an overview of the setup used to detect and to correct for the mismatch errors in a time-interleaved ADC. A digital signal generator is used to generate a deterministic and periodic test signal $r[n]$, that is applied in the form of $s[n]$ to the ADC by means of a DAC. The ADC is a p-channel time-interleaved ADC (as in Fig. 9.1), of which the gain, offset and time-skew can be adjusted by means of digitally controllable analog parameters. The channel outputs of the ADC are called $u_i[n]$ (with $1 \leq i \leq p$). A digital processing block is used to estimate the individual mismatches, and to control the analog parameters in order to minimize these errors iteratively. This section describes the test-signal generation and the error modelling. In the subsequent sections, the error detection and correction will be discussed.

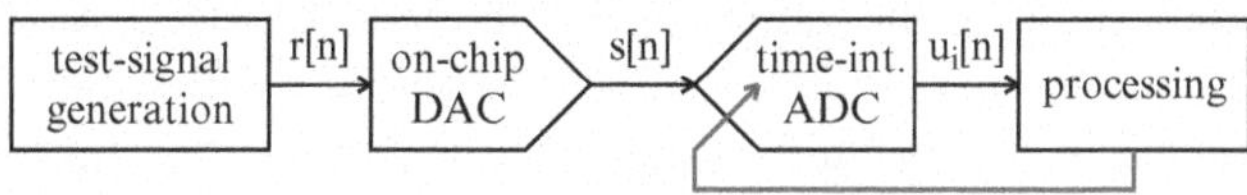

Figure 9.3. Detection and correction of channel mismatch errors.

3.2 Test-signal generation

In the presented setup, an on-chip signal generator is used to provide the ADC with a test-signal. By observing the resulting digital response of the ADC, the detection algorithm is able to determine estimations of the offsets, gain-errors and time-skews of the different channels of the ADC. Various input signals (sinusoids, white noise, etc.) could be used as an input signal within the given setup, but here a specific choice has been made to base the input signal on a pseudo-random maximum-length sequence (MLS) [84], because of several (non-exclusive) beneficial properties:

- An MLS has a wide frequency spectrum, up to the Nyquist frequency. As the signal contains high-frequency components, the response will be sensitive to time-skew errors. Moreover, the wide spectrum prevents that the system will be optimized for one specific frequency only, which would be the case with a single sinusoidal input signal.

- An MLS is always periodic; therefore, averaging of multiple measurements is possible if necessary. Obviously, averaging will come at the cost of a longer measurement time.

- An MLS can be used such that one can ensure that a large variety of input levels will be applied to the ADC, instead of applying only a few input levels. By doing so, local non-linearities in the transfer curve of the ADC can be averaged out.

- The hardware implementation of an MLS generator is simple, requiring only a few exclusive-OR gates and a number of flip-flops.

A 1-bit MLS signal $r[n]$ with a white frequency spectrum (where n denotes the sample moment) drives a 16-bit serial-in parallel-out shift register, which in turn drives an on-chip 16-bit binary-scaled DAC.[2] The analog output $s[n]$ of the DAC is used as the analog input signal for the ADC during the measurement phase. As the DAC will be used as well for other calibration methods, presented before in Chapters 5 and 8, it was designed as a 16-bit binary-scaled current-steering DAC. Despite the high resolution, the intrinsic accuracy of the DAC is less than 6-bit, making a simple and small implementation feasible. Figure 9.4 shows a functional diagram of the shift register and the DAC, where the α_i-parameters indicate the values of the elements of the DAC. By approximation, these parameters equal $\alpha_i = 2^{i-16}$, where the full scale range of the DAC is normalized to ± 1. From the diagram, one can see that the shift-register

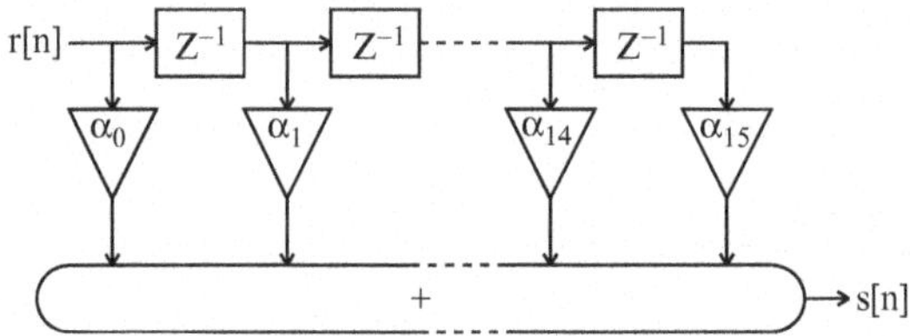

Figure 9.4. Model of the test signal generator, including the shift register and the DAC.

and DAC act as a filter $H_1(z)$ on signal $r[n]$:

$$H_1(z) = \sum_{i=0}^{15} \alpha_i z^{-i}\,, \tag{9.3}$$

thus, the analog output $s[n]$ is a filtered version of the 1-bit MLS $r[n]$. As $s[n]$ is constructed by multiple bits of $r[n]$, $s[n]$ can be treated as a *multi-bit* MLS. As such, the advantageous properties of the MLS input-signal remain valid for the DAC output-signal $s[n]$ as well. Figure 9.5 shows for both $r[n]$ and $s[n]$ the frequency spectrum and the discrete autocorrelation function. Note that the MLS order M equals 6 in these examples, resulting in a sequence length m of:

$$m = 2^M - 1 = 63 \tag{9.4}$$

The pseudo-random sequence $r[n]$ reaches a maximum autocorrelation for a shift of $p = 0$ only. As the DAC is composed of 16 elements, each controlled by a delayed version of $r[n]$ (as shown in Fig. 9.4), $s[n]$ reveals correlation for shifts up to $p = \pm 15$,[3] assuming no correlation ("white") in the MLS $r[n]$. However, because of the binary-scaled nature of the DAC, the correlation shows a steep roll-off and most of the correlation is contained closely around $p = 0$.

3.3 Channel mismatch model

In this subsection, the three mismatch errors (offset, gain error and time-skew) will be modelled. In order to model the effect of time-skew, an estimation of the transient behavior of the input signal has to be made. Here, it is assumed that the response is given by a single-pole system resulting in an exponential

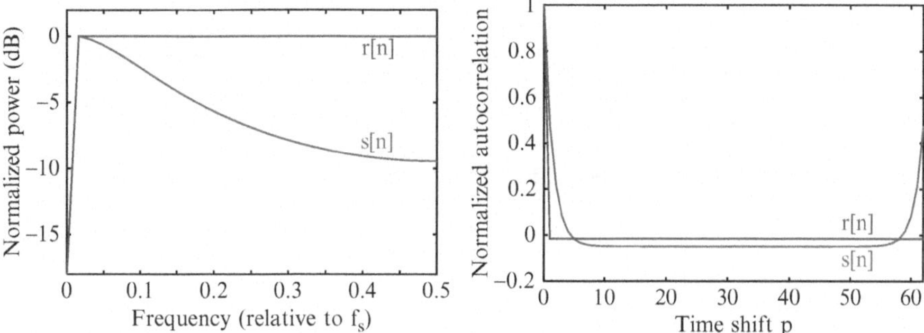

Figure 9.5. Normalized power spectrum (left) and normalized discrete autocorrelation function (right) of $r[n]$ and $s[n]$.

settling behavior, see Fig. 9.6. This assumption is valid as long as there is one dominant pole in the system, which could be for example the RC constant of the sampling capacitor of the T&H or the time-constant of the switch-drivers driving the DAC. In the implementation discussed later in this chapter, the dominant pole is given by the DAC driver, as the T&H was designed for a much higher frequency of operation than the DAC. Though the presented work is not limited to single-pole situations with a linear settling behavior, for simplicity of analysis this behavior was assumed. Figure 9.6 shows the settling behavior of the DAC, when the transition from level $s[n-1]$ to level $s[n]$ takes place. The behavior can be expressed in the time domain as:

$$s(t) = s[n] + e^{-t/\tau_d}(s[n-1] - s[n]), \tag{9.5}$$

where t is the time since the start of the transient and τ_d is the time-constant of the dominant pole. The sampled value $u'[n]$ of the T&H equals $s(t)$ taken at $t = t_{sample}$, where t_{sample} denotes the sample moment:

$$\begin{aligned} u'[n] &= s[n] + e^{-t_{sample}/\tau_d}(s[n-1] - s[n]) && (9.6)\\ &= (1-\beta)s[n] + \beta s[n-1], \text{ with: } \quad \beta = e^{-t_{sample}/\tau_d} && (9.7) \end{aligned}$$

Time-skew of the sample moment (i.e. a constant deviation in time) results in a slightly different value to be sampled. From (9.7), it follows that the time-skew error can be expressed as a filter operation, namely:

$$H_2(z) = (1 - \beta) + \beta z^{-1} \tag{9.8}$$

Note that for this specific model, $|\frac{\partial \beta}{\partial t_{sample}}|$ is maximum for $t_{sample} = 0$. Therefore, the highest time-skew sensitivity is achieved for $t_{sample} \downarrow 0$.

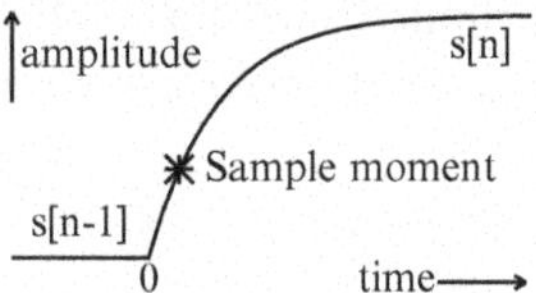

Figure 9.6. Transient behavior of the output of the DAC.

The gain and offset can be modeled as follows:

$$u[n] = \gamma u'[n] + \epsilon, \tag{9.9}$$

where γ denotes the gain and ϵ the offset of the channel. Finally, the diagram in Fig. 9.7 shows the three error mechanisms together. Note that t_{sample}, β, γ, ϵ and $u[n]$ will be different for each channel. Therefore, an index i will be added to these parameters in the remainder of this chapter.

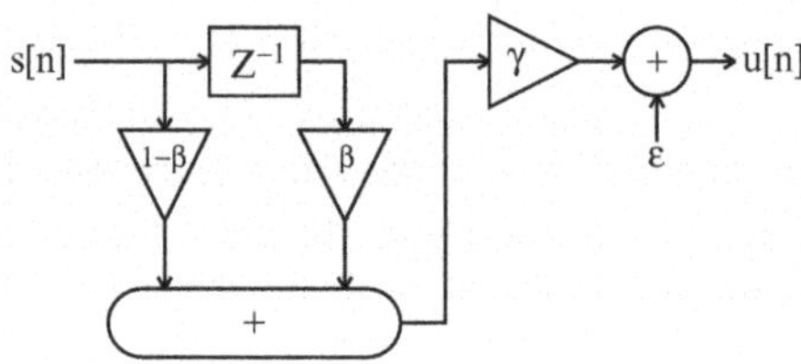

Figure 9.7. Model of a channel of a time-interleaved ADC, including offset, gain error and time-skew.

4. Channel mismatch detection

In the previous section, it was described how a periodic MLS $r[n]$ of length m was applied to a DAC to generate an analog test-signal $s[n]$ for the ADC. When the sequence-length m is chosen relatively prime to the number of channels p, after $m \cdot p$ sample moments, each value of the sequence $s[n]$ will have been applied exactly once to each channel of the ADC. An example for $p = 4$ and $m = 7$ is shown in Fig. 9.8: the seven samples of the sequence are numbered $1, \cdots, 7$. The first sample of the sequence is sampled by the first channel of the ADC. Then, the next sample (number 2) will be processed by the next channel (channel 2). After seven samples, the sequence will start again at sample 1. Likewise, after using the fourth channel, the first channel will take the next input signal. After $m \cdot p = 28$ sample moments, each of the seven input samples is applied exactly once to each channel. Furthermore, it can be noted that each individual channel receives the input stream in the same order, namely: $1, 5, 2, 6, 3, 7, 4$. As this pattern is known beforehand, the output data can be reordered to the original order of the samples $(1, 2, 3, 4, 5, 6, 7)$ for each individual channel. By doing so, the response $u_i[n]$ of each individual channel i of the ADC to the input signal $s[n]$ can be determined. Note that $u_i[n]$ approximates $s[n]$; for a mismatch-free channel, $u_i[n]$ will be equal to $s[n]$, and thus they will have the same frequency spectrum and autocorrelation function (Fig. 9.5). As each response $u_i[n]$ can be measured separately, the channel-mismatch information now also becomes available for each channel separately, which simplifies the detection algorithm that will be discussed later.

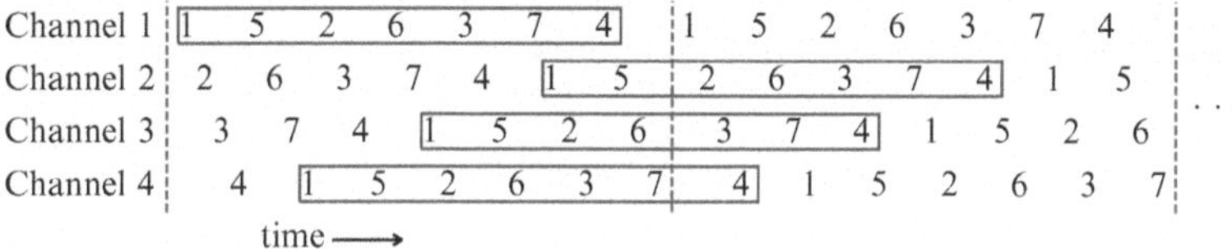

Figure 9.8. Channel responses for a 4-channel ADC with an input sequence length of 7.

The goal of the mismatch detection algorithm is to find for each ADC channel i estimations for the offset, gain and time-skew error (denoted by $\widehat{O}_{e,i}$, $\widehat{G}_{e,i}$ and $\widehat{\Delta}_i$, respectively), based on the measured output responses $u_i[n]$, such that the feedback algorithm can optimize the overall performance iteratively. Because of the iterative procedure, absolute accuracy of the estimations is not of extreme importance. Note that the length of each response equals the period length m of the original input signal $s[n]$, and that for the detection algorithm, $s[n]$ and the exact waveform produced by the DAC (including mismatches and dynamic behavior) are unknown signals. Nevertheless, correct estimations can be made by comparing the responses $u_i[n]$ with each other. First of all, a single reference channel is chosen arbitrarily (e.g. $i = 1$), against which the other channels will be compared. This means that $\widehat{O}_{e,i}$, $\widehat{G}_{e,i}$ and $\widehat{\Delta}_i$ will be determined and calibrated relative to channel 1, which is sufficient to minimize the mismatches. In the following, each of the three errors will be discussed.

4.1 Offset detection

The output offset of each channel of the ADC is composed of two components: a component which is common for all channels due to the generated input signal and imperfections in the DAC, and a second component which is different for each channel due to imperfections in the T&H and ADC. By simply subtracting the average value of $u_1[n]$ from the average value of $u_i[n]$, the common component will be cancelled out, and the relative offset value of channel i compared to channel 1 remains:

$$\widehat{O}_{e,i} = \frac{1}{m}\left(\sum_{n=0}^{m-1} u_i[n] - \sum_{n=0}^{m-1} u_1[n]\right) \tag{9.10}$$

The accuracy of the offset estimation process is dependent on the number of bits in the ADC and the length of the MLS. Assuming an N-bit ADC, with a full-scale range of $\pm FS$, the LSB of the converter equals:

$$LSB = \frac{2FS}{2^N} \tag{9.11}$$

From [30], it is known that the standard deviation of the quantization error will equal $\sqrt{\frac{1}{12}LSB^2}$. Assuming an equal amount of thermal noise, the total amount of noise added by the ADC becomes:

$$\sigma_{ADC} = \sqrt{\frac{1}{6}LSB^2} \tag{9.12}$$

The offset estimation in (9.10) adds up a total of $2m$ values, and divides by m, such that the standard deviation of the offset estimation becomes:

$$\sigma_{\widehat{O}_{e,i}} = \frac{1}{m} \cdot \sqrt{2m} \cdot \sigma_{ADC} = \sqrt{\frac{1}{3m}} \cdot LSB \approx \sqrt{\frac{1}{3 \cdot 2^M}} \cdot LSB \tag{9.13}$$

From this result, it becomes clear that the offset estimation can be more accurate than the accuracy of the ADC. Therefore, it is possible to achieve a precision beyond the requirement given in (9.2). Moreover, the measurement accuracy can be improved further by increasing the MLS order M. The model derived in (9.13) was verified by simulations on a two-channel ADC using various number-of-bits in the ADC and various MLS orders. Offsets between 0 and 10 mV were added to one channel of the ADC with a full-scale range of $\pm$1V, and the actual $\sigma_{\widehat{O}_{e,i}}$ was derived from the simulations. Figure 9.9 shows a good matching between the simulated and calculated results.

4.2 Gain error detection

The relative gain of channel i compared to channel 1 can be estimated by comparing the energy of the two responses:

$$\widehat{G}_{e,i} = \sqrt{\sum_{n=0}^{m-1} \left(u_i[n]\right)^2 \Big/ \sum_{n=0}^{m-1} \left(u_1[n]\right)^2} \tag{9.14}$$

To calculate the accuracy of the gain estimation, the expectation and standard deviation of $\sum_{n=0}^{m-1}\left(u_i[n]\right)^2$ are derived first: the quantized value $u_i[n]$ is rewritten as the sum of the analog value $u_i^*[n]$ and the ADC-noise error $q_i[n]$, and subsequently the squared term of $q_i[n]$ is neglected:

$$\sum_{n=0}^{m-1}\left(u_i[n]\right)^2 = \sum_{n=0}^{m-1}\left(u_i^*[n]+q_i[n]\right)^2 \approx \sum_{n=0}^{m-1}\left(u_i^*[n]\right)^2 + 2\sum_{n=0}^{m-1}\left(q_i[n]u_i^*[n]\right) \tag{9.15}$$

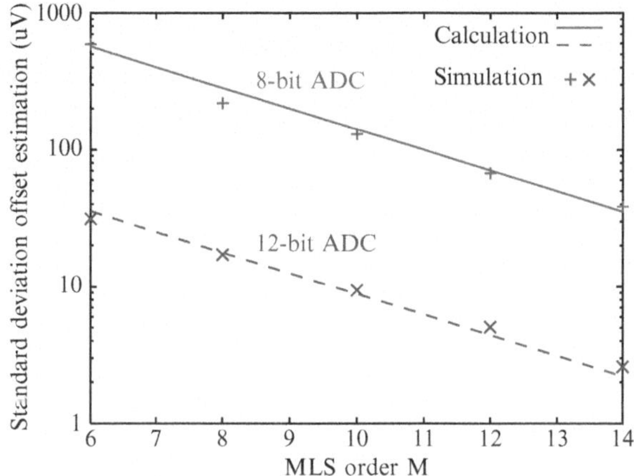

Figure 9.9. Standard deviation of the offset estimation as a function of the MLS order M, and the number of bits in the ADC.

The first term in this equation indicates the expected energy of the signal, and the second term expresses the random measurement error. As before (equation (9.12)), the standard deviation of the ADC noise equals $\sqrt{\frac{1}{6}LSB^2}$. Next, it can be shown that the rms-value of the multi-bit MLS $u_i^*[n]$ approximates $\sqrt{\frac{1}{3}FS^2}$, with FS the full-scale range (if the gain is close to unity). From that, it follows that the expected energy of (9.15) equals $\frac{m}{3}FS^2$ and the standard deviation of the random measurement error equals $2\sqrt{m}\sqrt{\frac{1}{6}LSB^2}\sqrt{\frac{1}{3}FS^2}$. With these results, (9.15) can be rewritten as:

$$E_i \cdot (1+\delta_i), \text{ with } E_i = \frac{m}{3}FS^2 \text{ and } \sigma_{\delta_i} = \sqrt{\frac{8}{m}} \cdot \frac{1}{2^N}, \tag{9.16}$$

where E_i is the expected energy of the signal, δ_i is the relative random measurement error with standard deviation σ_{δ_i}, and N is the number of bits in the

ADC. Using (9.16), (9.14) can be rewritten as:

$$\widehat{G}_{e,i} = \sqrt{\frac{E_i(1+\delta_i)}{E_1(1+\delta_1)}} = \sqrt{\frac{1+\delta_i}{1+\delta_1}} \approx \sqrt{1+\delta_i-\delta_1} \approx 1+\frac{1}{2}(\delta_i-\delta_1), \quad (9.17)$$

from which the standard deviation of the gain error estimation can be derived, yielding:

$$\sigma_{\widehat{G}_{e,i}} = \frac{1}{2}\sqrt{\sigma_{\delta_i}^2+\sigma_{\delta_1}^2} = \frac{\sigma_{\delta_i}}{\sqrt{2}} = \frac{2}{\sqrt{m}} \cdot \frac{1}{2^N} \approx \frac{2}{\sqrt{2^M}} \cdot \frac{1}{2^N} \quad (9.18)$$

This result shows, as previously with the offset estimation, that the gain estimation can be more accurate than the accuracy of the ADC, and meets the accuracy requirement given in (9.2). Also, the accuracy can be improved further by increasing the MLS order M. Again, simulations on a two-channel ADC were performed to verify the calculations for various MLS orders and various number-of-bits in the ADC. The results in Fig. 9.10 confirm the presented model.

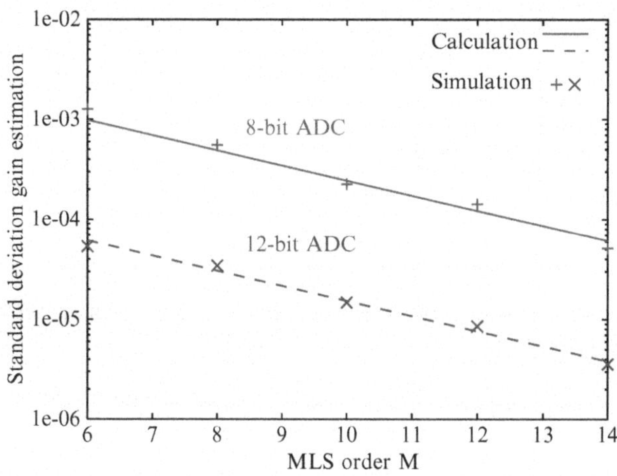

Figure 9.10. Standard deviation of the gain estimation as a function of the MLS order M, and the number of bits in the ADC.

4.3 Time-skew detection

For the estimation of the time-skew error, we use the property that the output signal of the DAC is a time-continuous signal, with a certain settling behavior just after the code transitions as explained before in Fig. 9.6. For clarity, an example is shown in Fig. 9.11: a multi-bit MLS sequence of length 3 is sampled by a 2-channel converter. The MLS samples are indicated by $s[1]$, $s[2]$, and $s[3]$. After these three samples, the sequence is repeated as indicated in the figure. The equivalent analog amplitude of the digital DAC input signal takes one out of three values, corresponding to the applied MLS signal. The analog output of the DAC, which is also the input of the ADC, approximates the DAC input level. However, the analog signal shows a settling behavior just after the code transitions due to limited bandwidth. The sample instants of the AD converter are set to occur during the code transitions. Because of that, time-skew in the sampling clock will affect the sampled values, as the input signal is time-dependent around the sampling moment. The figure shows the sampling instants for the two channels of the converter, while a time-skew is present between the two channels. After six sampling moments, both channels have sampled the three different levels of the input signal. From these six samples, the response of each individual channel to the input sequence of length 3 can be determined: the first period of the MLS sequence (between sample time 0 and 3) yields the response of channel 1 to two of the three MLS levels. The response to the third value of the MLS sequence can be taken from the second period of the MLS sequence (between sample time 3 and 6). Likewise for the second channel, the first MLS period yields one value of the response, while the second MLS period yields the remaining two values of the response.

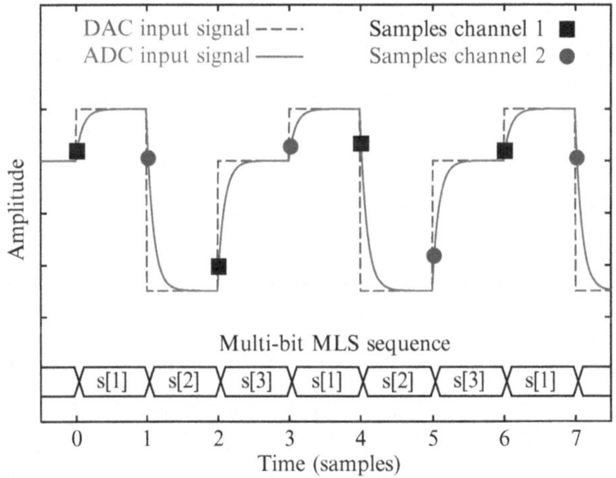

Figure 9.11. Time-domain behavior of a 2-channel converter with a multi-bit MLS sequence of length 3.

By plotting the two MLS periods on top of each other (Fig. 9.12), the two responses can be compared visually. From this figure, it becomes clear that channel 2 is lagging behind (because of time-skew), resulting in a different channel response compared to channel 1. Thus, a time-skew error Δ_i of an ADC channel results in a change of the measured response $u_i[n]$. In the following, two methods will be presented to estimate the time-skew based on the change in $u_i[n]$: the first method is based on cross-correlation and the second on frequency-domain analysis.

Time-skew detection based on cross-correlation

$R_{1,i}[p]$ is defined as the discrete cross-correlation between the response of channel 1 and the response of channel i, where the offset and gain of channel i are already corrected using the previously estimated values:

$$R_{1,i}[p] = \sum_{n=0}^{m-1} u_1[n] \cdot \frac{u_i[n-p] - \widehat{O}_{e,i}}{\widehat{G}_{e,i}} \tag{9.19}$$

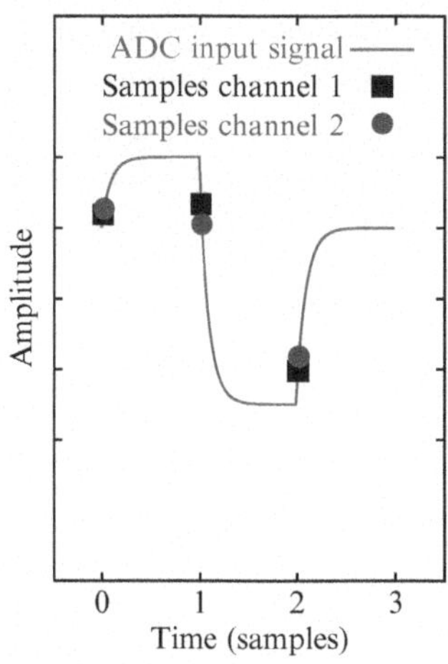

Figure 9.12. Effective time-domain response for the individual channels of a time-interleaved converter with a multi-bit MLS sequence of length 3.

In case channel i is perfectly matched, $R_{1,i}[p]$ will be equal to the discrete autocorrelation of channel 1: $R_{1,1}[p]$. Hence, the difference $D_i[p]$ will be related to the time-skew error Δ_i:

$$D_i[p] = R_{1,i}[p] - R_{1,1}[p] \tag{9.20}$$

Neglecting gain and offset for the moment, (9.7) gives:

$$u_i[n] = (1 - \beta_i)s[n] + \beta_i s[n-1] \tag{9.21}$$

Assuming that reference channel 1 samples at $t = t_{sample,nom}$ and channel i has a time-skew Δ_i, one can rewrite β_i as follows using (9.7):

$$\beta_1 = e^{-t_{sample,nom}/\tau_d} \tag{9.22}$$

$$\beta_i = e^{-(t_{sample,nom}+\Delta_i)/\tau_d} = e^{-t_{sample,nom}/\tau_d} \cdot e^{-\Delta_i/\tau_d} \approx \beta_1 \cdot (1 - \frac{\Delta_i}{\tau_d}) \quad (9.23)$$

As explained before, the highest sensitivity for the time-skew detection is reached during the steepest part of the code transition (see Fig. 9.6), thus for $t_{sample,nom} \downarrow 0$. In that case, β_1 will be close to one (according to (9.22)). As such, (9.21) can be rewritten as:

$$u_i[n] \approx u_1[n] + \frac{\Delta_i}{\tau_d}(s[n] - s[n-1]) \quad (9.24)$$

Substituting (9.24) in (9.19) and neglecting gain and offset yields:

$$R_{1,i}[p] \approx R_{1,1}[p] + \frac{\Delta_i}{\tau_d} \sum_{n=0}^{m-1} u_1[n] \cdot (s[n-p] - s[n-p-1]) \quad (9.25)$$

As $\beta_1 \approx 1$, $u_1[n]$ approximates $s[n-1]$, thus combining (9.20) with (9.25) results in:

$$\begin{aligned} D_i[p] &\approx \frac{\Delta_i}{\tau_d} \sum_{n=0}^{m-1} s[n-1] \cdot (s[n-p] - s[n-p-1]) \\ &= \frac{\Delta_i}{\tau_d}(R_s[p-1] - R_s[p]), \end{aligned} \quad (9.26)$$

with $R_s[p]$ the discrete autocorrelation of $s[n]$, as illustrated in Fig. 9.5. From (9.26) it follows that $D_i[p]$ is proportional to the time-skew Δ_i, and thus $D_i[p]$ can be used to estimate the time-skew based on the measured channel responses. Moreover, as explained previously and as illustrated in Fig. 9.5, it can be observed that most of the information of $R_s[p]$ is contained around $p = 0$, up to $p = 15$. Therefore, a partial sum over $D_i[p]$ is taken that covers the relevant values from $R_s[p]$ ($p = 0 \cdots 16$):

$$\sum_{p=1}^{16} D_i[p] \approx \frac{\Delta_i}{\tau_d}(R_s[0] - R_s[16]) \quad (9.27)$$

Corresponding to Fig. 9.5, it can be shown that:

$$R_s[0] \approx m \sum_{i=0}^{15} \alpha_i^2 \approx \frac{m}{3} \text{ and } R_s[16] \approx 0 \Rightarrow \sum_{p=1}^{16} D_i[p] \approx \frac{\Delta_i}{\tau_d}\frac{m}{3}, \quad (9.28)$$

such that the time-skew Δ_i can be estimated by:

$$\widehat{\Delta}_i = \frac{3\tau_d}{m} \sum_{p=1}^{16} D_i[p] \quad (9.29)$$

The behavior of $D_i[p]$ as a function of the time-skew Δ_i was verified with a transistor-level simulation of a DAC and a 2-channel T&H. The sample frequency of the DAC and the T&H was set to $f_s = 10$ MSps, and various time-skews ($\pm$1ps, $\pm$10 ps and $\pm$100 ps) were added to channel 2 of the T&H. Figure 9.13 shows both the results for $D_2[p]$, based on the simulated responses $u_i[n]$, and the resulting estimation $\widehat{\Delta}_i$ as a function of Δ_i, based on (9.29). Figure 9.13 (left) shows that the amplitude of $D_2[p]$ is proportional to the applied time-skew, as expected based on (9.26). Also, it can be seen that the highest time-skew sensitivity is achieved for small values of p, which corresponds to the fact that $R_s[p]$ (Fig. 9.5) contains most of the information in that region. The estimation $\widehat{\Delta}_i$ is proportional to Δ_i, but slightly smaller in value. This is because in theory, sampling takes place at the steepest part of the transition: $t_{sample,nom} \downarrow 0$. In practice, sampling takes place slightly later, at a less steep part of the transition. Because of that, the estimated time-skew will be underestimated. Nonetheless, as the estimation is still proportional to the actual time-skew, it provides sufficient information for the feedback algorithm to optimize the performance.

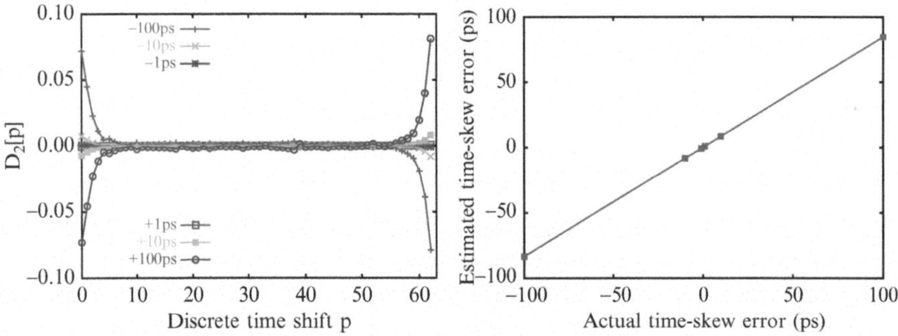

Figure 9.13. $D_2[p]$ for various time-skews ($\pm$1, $\pm$10 and $\pm$100 ps) (left), and the estimated time-skew (right).

Calculating the accuracy of the time-skew estimation is not so straightforward: (9.29) shows that $\sigma_{\widehat{\Delta}_i}$ will be equal to $\frac{3\tau_d}{m}$ times the standard deviation of the summation over $D_i[p]$. Ultimately, $D_i[p]$ is based on summations of m observations of $u_i[n]$. Assuming ADC noise as before (equations (9.11) and (9.12)), the standard deviation of $D_i[p]$ will be proportional to $\sqrt{m}/2^N$, thus:

$$\sigma_{\widehat{\Delta}_i} \propto \frac{3\tau_d}{m} \frac{\sqrt{m}}{2^N} \propto \frac{\tau_d}{\sqrt{2^M} 2^N} \tag{9.30}$$

The validity of this equation will be verified later in this chapter (see page 144).

Time-skew detection based on frequency-domain analysis

As an alternative to cross-correlation in the time-domain, frequency analysis in the Z-domain can also be used to estimate the time-skew. Previously, the transfer functions of the DAC and the ADC, including time-skew, were modeled as $H_1(z)$ and $H_2(z)$, respectively. Thus, the output of channel i of the ADC can be written as follows in the Z-domain:

$$U_i(z) = H_1(z) \cdot H_{2,i}(z) \cdot R(z), \tag{9.31}$$

where $R(z)$ is the Z-domain equivalent of the MLS sequence $r[n]$. Note that $H_1(z)$ and $R(z)$ are common for all channels, but $H_{2,i}(z)$ (representing the time-skew) is different for each channel i. The relative frequency response of a channel now becomes:

$$\frac{U_i(z)}{U_1(z)} = \frac{H_{2,i}(z)}{H_{2,1}(z)} = \frac{1 - \beta_i + \beta_i z^{-1}}{1 - \beta_1 + \beta_1 z^{-1}} \tag{9.32}$$

Using the same approximations for β_1 and β_i as before, this can be estimated as follows:

$$\begin{aligned} \frac{U_i(z)}{U_1(z)} &= \frac{1 - \beta_i + \beta_i z^{-1}}{1 - \beta_1 + \beta_1 z^{-1}} \approx \frac{1 - \beta_1 \cdot (1 - \frac{\Delta_i}{\tau_d}) + \beta_1 \cdot (1 - \frac{\Delta_i}{\tau_d}) z^{-1}}{1 - \beta_1 + \beta_1 z^{-1}} \\ &\approx \frac{\beta_1 \cdot \frac{\Delta_i}{\tau_d} + \beta_1 \cdot (1 - \frac{\Delta_i}{\tau_d}) z^{-1}}{\beta_1 z^{-1}} = 1 - \frac{\Delta_i}{\tau_d} \cdot (1 - z) \end{aligned} \tag{9.33}$$

As $u_i[n]$ is a discrete-time signal, a discrete fourier transformation (DFT) can be used to transform the measured values $u_i[n]$ to $U_i[k]$. As $u_i[n]$ is periodic with length m, $U_i[k]$ is defined for $0 \leq k < m$. $U_i[k]$ corresponds to a sampled version of $U_i(z)$ such that:

$$U_i[k] = U_i(z)\Big|_{z=e^{j2\pi k/m}} \tag{9.34}$$

Substituting (9.34) in (9.33) yields the following relation between the measured responses $U_i[k]$ and the time-skew Δ_i:

$$\frac{U_i[k]}{U_1[k]} \approx 1 - \frac{\Delta_i}{\tau_d} \cdot (1 - e^{j2\pi k/m}) \tag{9.35}$$

For illustration, Fig. 9.14 shows this function in the complex domain for $\Delta_i/\tau_d = 0.01$ and for $\Delta_i/\tau_d = 0.02$. Equation (9.35) has two main components: a relatively large constant value (1) and a circle, of which the diameter is proportional to the time skew Δ_i. When the terms of (9.35) are multiplied by $e^{-j2\pi k/m}$ and a summation is taken, the constant value is canceled, while the components describing the circle add up:

$$\sum_{k=0}^{m-1} e^{-j2\pi k/m} \frac{U_i[k]}{U_1[k]} \approx \sum_{k=0}^{m-1} e^{-j2\pi k/m} \cdot \left(1 - \frac{\Delta_i}{\tau_d}\right) + \frac{\Delta_i}{\tau_d} \tag{9.36}$$

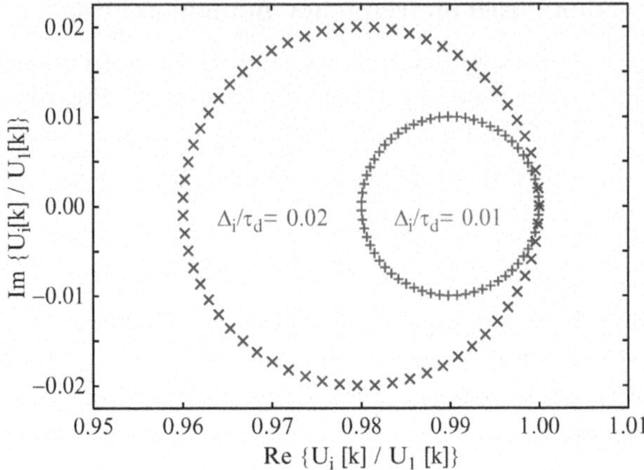

Figure 9.14. Expected $U_i[k]/U_1[k]$ for $m = 63$ and various Δ_i/τ_d.

As the summation over $e^{-j2\pi k/m}$ reduces to zero, (9.36) simplifies to:

$$\sum_{k=0}^{m-1} e^{-j2\pi k/m} \cdot \left(1 - \frac{\Delta_i}{\tau_d}\right) + \frac{\Delta_i}{\tau_d} = \sum_{k=0}^{m-1} \frac{\Delta_i}{\tau_d} = m\frac{\Delta_i}{\tau_d}, \tag{9.37}$$

and therefore, the time-skew Δ_i can be estimated by:

$$\widehat{\Delta}_i = \frac{\tau_d}{m} \sum_{k=0}^{m-1} e^{-j2\pi k/m} \frac{U_i[k]}{U_1[k]} \tag{9.38}$$

For visualization, it is more practical to use the absolute value of $U_i[k]/U_1[k]$ as a function of k (i.e. the discrete frequency spectrum) instead of the complex value of $U_i[k]/U_1[k]$. The absolute value of (9.35) can be approximated as follows:

$$\left|\frac{U_i[k]}{U_1[k]}\right| \approx \left|1 - \frac{\Delta_i}{\tau_d} \cdot (1 - e^{j2\pi k/m})\right| = \left|1 + \frac{\Delta_i}{\tau_d}\left(\cos(\frac{2\pi k}{m}) - 1\right)\right.$$
$$\left. + j\frac{\Delta_i}{\tau_d}\sin(\frac{2\pi k}{m})\right| \tag{9.39}$$

As the imaginary part of this equation is negligible compared to the real part, it simplifies to:

$$\left|\frac{U_i[k]}{U_1[k]}\right| \approx 1 + \frac{\Delta_i}{\tau_d} \cdot \left(\cos(\frac{2\pi k}{m}) - 1\right), \tag{9.40}$$

revealing that the relative frequency response is a raised-cosine, where the amplitude of the cosine term is dependent on the time-skew. Note that the cosine term corresponds to the real part of the circle, described in (9.35). For illustration, Fig. 9.15 shows the frequency response for various time-skews, using equation (9.35).

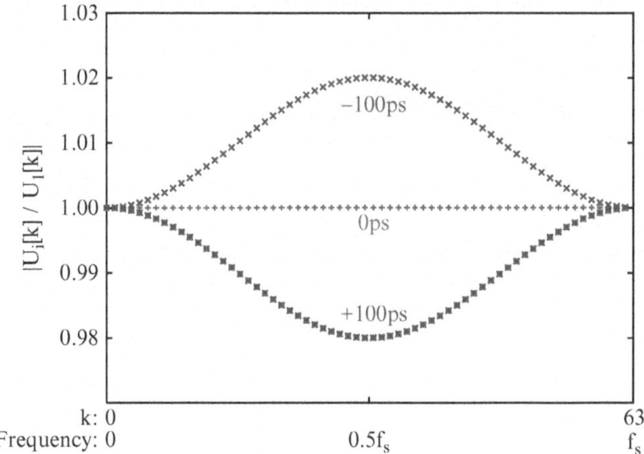

Figure 9.15. Expected $|U_i[k]/U_1[k]|$ for various time-skews (0, $\pm$100 ps) with $\tau_d = 10$ ns.

In the complex domain, the time-skew is estimated by multiplying the response $U_i[k]/U_1[k]$ by $e^{-j2\pi\frac{k}{m}}$, followed by a summation. As $|U_i[k]/U_1[k]|$ approximates the real part of $U_i[k]/U_1[k]$, multiplying $|U_i[k]/U_1[k]|$ by the real part of $e^{-j2\pi k/m}$ (i.e. $\cos(2\pi k/m)$), followed by a summation, results in a similar method to estimate the time-skew:

$$\begin{aligned}\sum_{k=0}^{m-1}\cos(\frac{2\pi k}{m})\left|\frac{U_i[k]}{U_1[k]}\right| &\approx \sum_{k=0}^{m-1}\cos(\frac{2\pi k}{m})\Big(1-\frac{\Delta_i}{\tau_d}+\frac{\Delta_i}{\tau_d}\cos(\frac{2\pi k}{m})\Big)\\ &= \sum_{k=0}^{m-1}\cos^2(\frac{2\pi k}{m})\frac{\Delta_i}{\tau_d} = \sum_{k=0}^{m-1}\Big(\frac{1}{2}+\frac{1}{2}\cos(\frac{4\pi k}{m})\Big)\frac{\Delta_i}{\tau_d}\\ &= \frac{m}{2}\cdot\frac{\Delta_i}{\tau_d} \qquad (9.41)\end{aligned}$$

Compared to the estimation in the complex domain (equation (9.37)), the summation in the real domain is a factor of two smaller. This is logical, as the imaginary part of the expression is neglected in the real domain.

The frequency-domain method for the estimation of the time-skew was also verified with transistor-level simulations. Using the same data as previously for the correlation-based estimation (Fig. 9.13), the results of Fig. 9.16 were obtained. As before with the cross-correlation method, the time-skew is underestimated but proportional to the actual error.

Comparison of the time-skew estimation methods

To compare the accuracy of both time-skew estimation methods, the transistor-level simulation results were used to determine the standard deviation of the estimations, based on the noise generated by the ADC according to (9.12). Figure 9.17 shows the results as a function of the number-of-bits in the ADC, revealing that both methods achieve a similar performance which can be approximated by the same trend-line.

Accuracy of the time-skew estimation

While Fig. 9.17 shows the estimation accuracy as a function of the number of bits in the ADC, Fig. 9.18 shows the effect of both the number of bits in the ADC and the MLS order M. For the results in Fig. 9.18, high-level simulations were performed using the frequency-domain time-skew estimation method, a

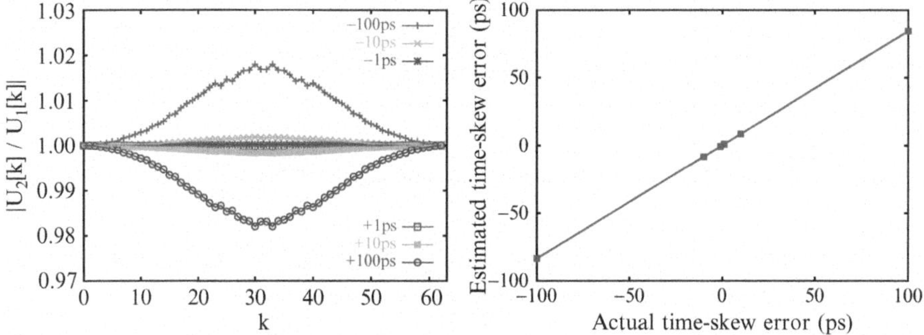

Figure 9.16. $|U_i[k]/U_1[k]|$ for various time-skews (±1, ±10 and ±100 ps) (left), and the estimated time-skew (right).

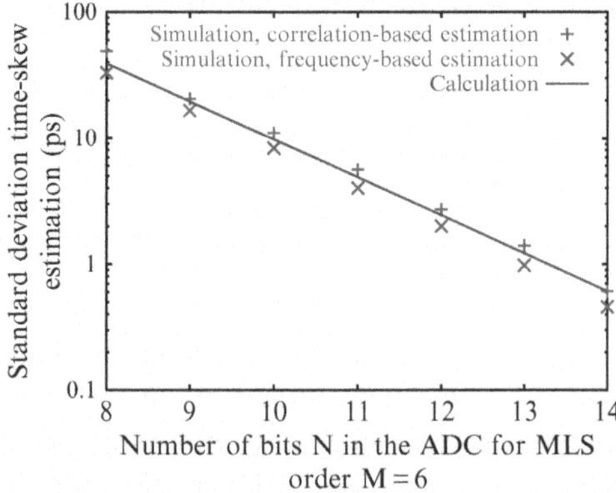

Figure 9.17. Standard deviation of the time-skew estimation for the two methods using transistor-level simulations. The calculation refers to equation (9.42).

time-constant τ_d of 20 ns and ADC noise according to (9.12). Corresponding to (9.30), the figure shows the relation between $\sigma_{\widehat{\Delta}_i}$, the MLS order M and the number of bits N, yielding the following empirical formula for the standard deviation of the time-skew estimation:

$$\sigma_{\widehat{\Delta}_i} \approx \frac{4}{\sqrt{2^M}} \cdot \frac{1}{2^N} \cdot \tau_d \tag{9.42}$$

For comparison, this estimation is also plotted in Figs. 9.17 and 9.18. As previously with the gain and offset estimations, this equation shows that also the time-skew estimation can be improved by increasing the MLS order M. Because of that, the self-measurement accuracy can be made far smaller than the time constant τ_d of the measurement setup. As such, the precision requirement of (9.2) can be reached without the need for a high-speed self-measurement DAC. Actually, the slewing of the DAC is used for the determination of the time-skew.

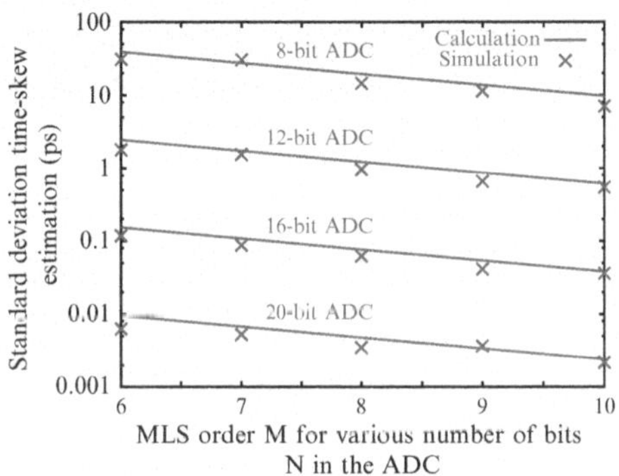

Figure 9.18. Standard deviation of the time-skew estimation using high-level simulations.

5. Channel mismatch correction

In the previous section, the methods to estimate the gain, offset and time-skew for each ADC channel were discussed. The next step is to actually correct the errors based on the error estimations. First, the actual implementation of the correction hardware will be discussed, and then the algorithm to control the correction circuit based on the error estimations will be described.

5.1 Analog correction method

In reality, the channel mismatches that were modeled by offset, gain error and time-skew, originate from many different error sources within the channel. For example, mismatch of the components in the T&H, ADC or clock circuitry, mismatch in the reference levels for the different channels, mismatch in the layout, gradients due to layout and gradients due to voltage drop. Even though the actual error sources might be unknown, they can be compensated together at a single location in the channel. In the presented solution, the actual correction takes place in the T&H circuit by tuning several analog components, such that no additional digital processing power is required for the correction. For the correction of gain and offset errors, the same implementation as in Chapter 8 will be used. On top of that, a programmable delay is added to the switch-driver to provide a correction mechanism for the time-skew. The implementation of the controllable T&H is visualized in Fig. 9.19, showing the three 8-bit programmable parameters: the two current sources I_a and I_b of the differential pair and the delay τ of the clock buffer.

As described in Chapter 8, the gain can be controlled by changing I_a and I_b in the same direction, while the offset can be controlled by changing I_a and I_b in opposite directions. The time-skew can be controlled by changing the delay τ of the switch-driver. The transistor-level implementation of the programmable delay will be reviewed later in Section 7. Figure 9.20 shows the controllability of the three parameters based on transistor-level simulations. The strong non-linear behavior of the time-skew correction is not directly

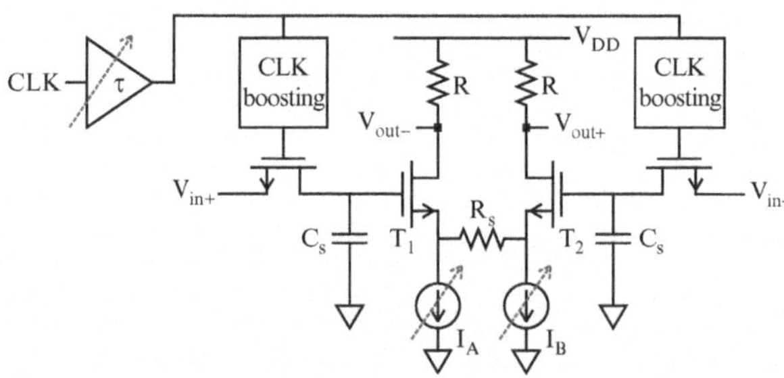

Figure 9.19. Implementation of the T&H with programmable gain, offset and time-skew.

important as the algorithm is able to find the optimum regardless of the non-linearity. However, the non-linearity also results in a variable step-size across the tuning range. As such, part of the curve needs to be over-designed in order to achieve sufficient performance throughout the complete range. The main cause of the strong non-linearity of this curve is due to the wide tuning range (250 ps). When redesigning for a more realistic range (e.g. 50 ps), the linearity will improve automatically. For all three parameters, the 8-bit control allows enough range to compensate worst-case mismatch situations while the step size is small enough to guarantee enough calibration accuracy to realize 10-bit post-calibration performance for a 2-channel time-interleaved ADC operating at 1 GSps.

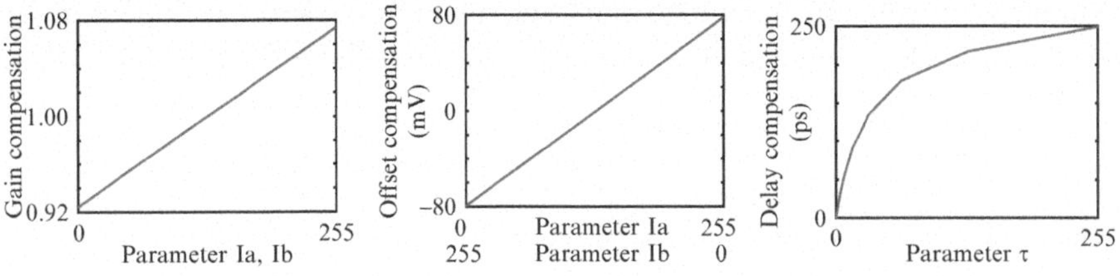

Figure 9.20. Analog controllability of gain, offset and time-skew.

5.2 Correction algorithm

The correction algorithm updates the parameter settings of each channel ($I_{a,i}$, $I_{b,i}$ and τ_i) based on the measured estimations of the errors ($\widehat{O}_{e,i}$, $\widehat{G}_{e,i}$ and $\widehat{\Delta}_i$). After the update, a new measurement will be performed and the parameters will be updated again until a situation with sufficient performance is achieved. Because of the iterative error-optimization procedure, the absolute accuracy of the estimations given by (9.10), (9.14) and (9.29) or (9.38) is not of extreme importance, as long as the feedback-loop controls the errors towards zero. For the same reasons, the actual (non-linear) behavior of the analog correction circuit is not critical. As in Chapter 8, the parameter-update functions are simple linear combinations of the error estimations:

$$\left\{ \begin{array}{rcrcl} I_{a,i}[k] & = & I_{a,i}[k-1] & - & c_1 \cdot (\widehat{G}_{e,i}[k]-1) - c_2 \cdot \widehat{O}_{e,i}[k] \\ I_{b,i}[k] & = & I_{b,i}[k-1] & - & c_1 \cdot (\widehat{G}_{e,i}[k]-1) + c_2 \cdot \widehat{O}_{e,i}[k] \\ \tau_i[k] & = & \tau_i[k-1] & + & c_3 \cdot \widehat{\Delta}_i[k] \end{array} \right. , \quad (9.43)$$

where c_1, c_2 and c_3 are constants, and k indicates the iteration of the feedback algorithm.

6. Simulation results

The calibration method was verified by means of transistor-level simulations on a 2-channel time-interleaved ADC, operating at $f_s = 1$ GSps. The DAC, clock circuitry and T&H circuits were implemented fully on transistor-level in a CMOS 0.18 μm technology. The quantizers were implemented with ideal 12-bit ADCs, having a full-scale range of $V_{pp} = 1$ V. The SFDR of a single channel is around 65 dB, limiting the overall performance to 10-bit at most. The following combination of mismatch errors was applied to one channel of the ADC: $O_{e,2} = 1.6$ mV, $G_{e,2} = 1.0064$ and $\Delta_2 = 30$ ps. This set of errors was chosen as they correspond to 8-bit performance while operating at $f_s = 100$ MSps, which is a realistic performance for an intrinsic circuit without calibration. Applying a sinusoid with frequency $\frac{347}{1024} f_s$ [4] yields the output spectrum of Fig. 9.21. The distortion components of the T&H (HD3 and HD5) can be seen as well as several components due to channel mismatch: a DC-level due to offset mismatch, and a tone due to gain and time-skew mismatch. In this case, the time-skew mismatch is dominant, limiting the SFDR of the ADC to 29.7 dB.

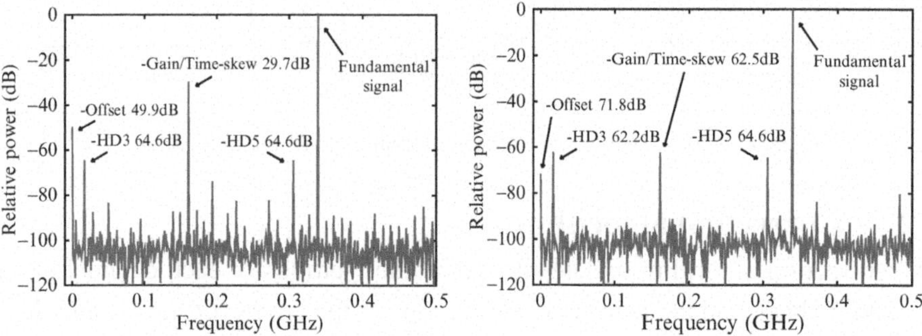

Figure 9.21. Output power spectrum for the 2-channel ADC, before calibration (left) and after calibration (right).

Next, the self-measurement and self-correction algorithms were applied to the ADC. For this purpose, a transistor-level implementation of a DAC (with an intrinsic accuracy of only 6-bit) was used as a test-signal generator. As the sample frequency of the DAC is limited to 100 MSps, during this calibration phase the time-interleaved ADC was also set to operate at $f_s = 100$ MSps. Also, due to simulation limitations the MLS-order was limited to $M = 6$. From equations (9.13), (9.18) and (9.42), the self-measurement accuracy can be estimated. Knowing that $M = 6$, $LSB = 0.24$ mV, $N = 12$, $\tau_d = 3$ ns, it follows that: $\sigma_{\widehat{O}_{e,i}} = 0.02$ mV, $\sigma_{\widehat{G}_{e,i}} = 0.006\%$ and $\widehat{\Delta}_i = 0.4$ ps. However, especially for gain and offset calibration, the post-calibration performance is limited by the step-size of the analog correction elements: 0.6 mV for the offset and 0.06% for the gain. Nonetheless, the self-measurement accuracy and the correction accuracy are sufficient to achieve 10-bit post-calibration performance.

After applying the calibration algorithm, the performance of the ADC was verified again. For this purpose, f_s was set back to the original value of 1 GSps. The output power spectrum after calibration is visualized in Fig. 9.21. It can be seen that the distortion components HD3 and HD5 of the T&H remain similar to the original performance. This is to be expected, as they are not influenced by the channel mismatches. However, some linearity variations are possible when the gain and offset are tuned, because of the open-loop T&H architecture. The spurious components due to offset and gain/time-skew were reduced to -71.8 and -62.5 dB, respectively. As a result, the SFDR of the ADC improves from 29.7 to 62.2 dB. The dominant spurious component after calibration is the HD3 component, which shows that the post-calibration accuracy is not limited by the channel mismatches anymore, but by the linearity of the channel itself. Based on the final performance, the post-correction matching accuracy can be derived, yielding $O_{e,2} = 0.08$ mV, $G_{e,2} = 1.0012$ and $\Delta_2 = 0.4$ ps.

Finally, the stability of the calibration as a function of the temperature of the environment was simulated. The transistor-level implementation which was previously calibrated at a temperature of 300 K was simulated at different temperatures without updating the calibration parameters. Figure 9.22 shows the effect on offset, gain-error and time-skew, showing that the performance degrades when the temperature deviates from the temperature during calibration. However, even within a temperature range of 100 degrees, the post-calibration performance remains around ten times better compared to the pre-calibration performance.

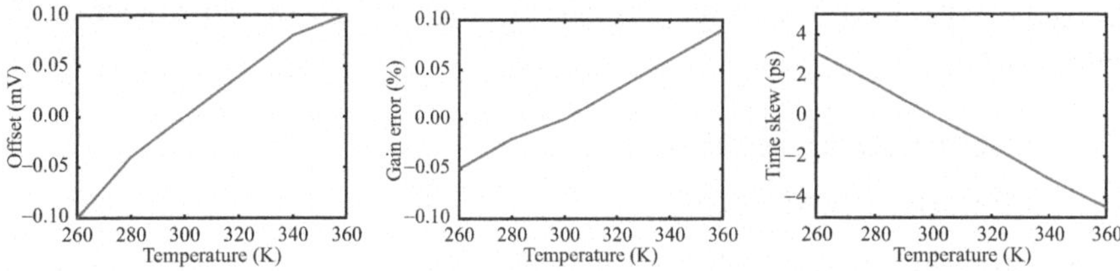

Figure 9.22. Offset, gain-error and time-skew as a function of the temperature.

7. Implementation of the calibration method and layout

For the experimental verification of the time-interleaved calibration method, the design from Section 7 was reused. This design already includes two T&H's with programmable gain and offset and a DAC for the generation of the test-signal. On top of that, a programmable delay element is added to the clock driver of each T&H to facilitate the time-skew correction. The implementation of the programmable delay element is shown in Fig. 9.23, where 8 CMOS inverters are connected in parallel. The $\frac{W}{L}$'s of these inverters are binary-scaled. Each of the eight inverters can be enabled or disabled by a control bit that controls the supply to each inverter. Enabling less or more inverters changes the RC-constant of the buffer, thereby affecting the delay of the clock driver. The realized delay, as a function of the 8-bit control signal is also shown in Fig. 9.23.

The added power consumption of the programmable inverter array varies between 0 and 7 mW, dependent on the selected delay. The relatively large consumption is caused by the large tuning range of 250 ps. For a smaller tuning range, e.g. 50 ps, the average power consumption of the programmable delay would be only 0.1 mW. The layout of the overall structure is shown in Fig. 9.24, including two T&H's, programmable current sources, programmable clock

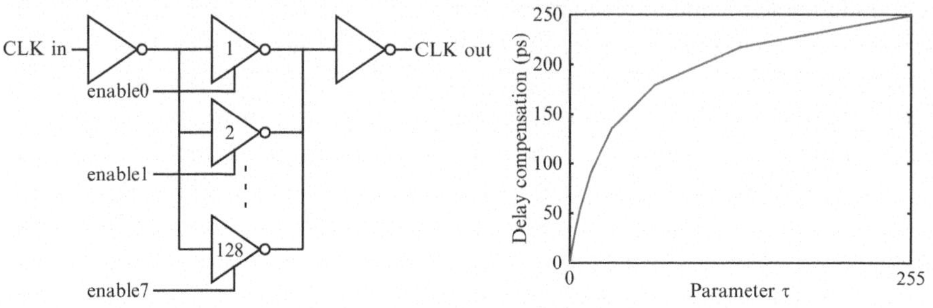

Figure 9.23. Programmable clock delay circuit (left) and its delay programmability (right).

drivers, digital logic and configuration switches. The configuration switches are used to make to connections between the various blocks, dependent on the desired setup.

8. Experimental results

8.1 Measurement setup

For the experimental verification of the calibration method, a dedicated measurement setup was created. For the self-measurement phase of the calibration method, the setup in Fig. 9.25 was implemented: a 10th-order MLS sequence is generated inside an FPGA and applied to the DAC that is available inside the

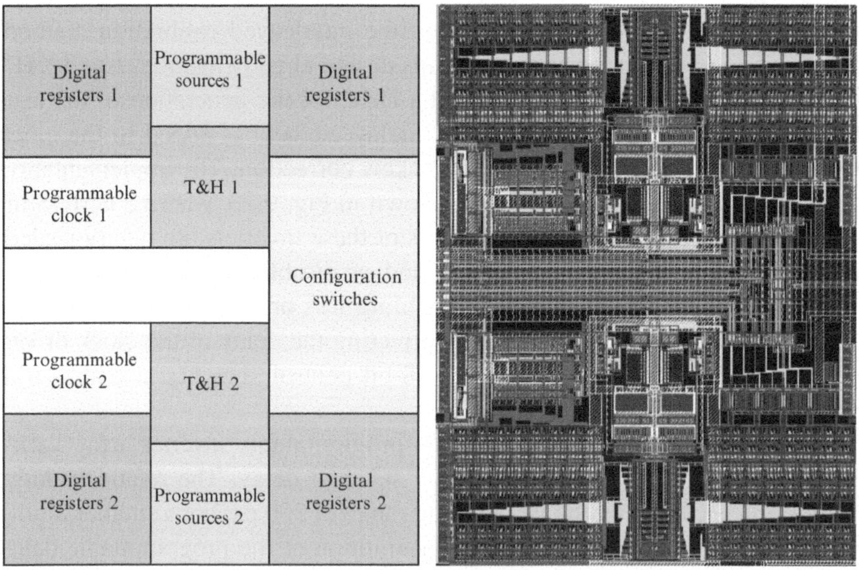

Figure 9.24. Floorplan and layout of the 2-channel T&H, measuring 370 × 280 μm.

test-chip. Even though the implemented DAC (Chapter 5) has a resolution of 16-bit, its accuracy is limited to 6 bits. Then, the analog test signal is applied to both T&H's. By means of a multiplexer, the two outputs of the T&H's are combined, sampled by an off-chip ADC and processed externally to run the calibration algorithm. Because the multiplexer is driven indirectly by reprogramming the entire configuration register inside the chip, the sampling rate is limited to 10 kSps. Despite the low sampling rate, a correct performance evaluation of the time-interleaved T&H is still possible, as the effect of the mismatches is not dependent on the sampling rate but only on the properties of the input signal. By applying a full-scale high-frequency input tone, the effect of the mismatches can be observed precisely.

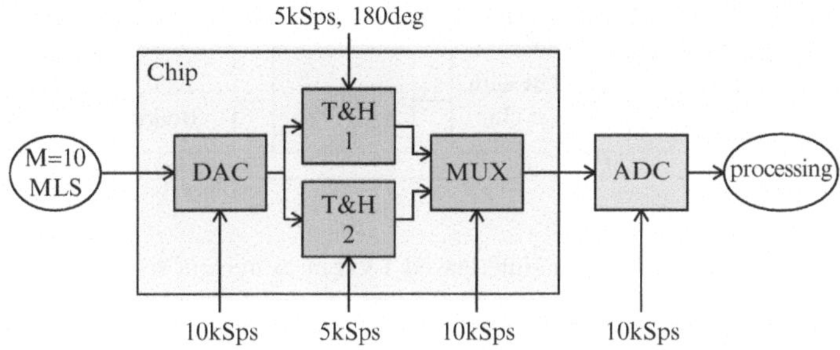

Figure 9.25. Setup for self-measurement of channel mismatches.

For the verification of the achieved performance of the T&H, both before and after calibration, the setup in Fig. 9.26 is used: an external sinusoid is now applied as an input. This signal is fed to the two T&H circuits, multiplexed, and sampled by the off-chip ADC. From the ADC output, the output spectrum of the time-interleaved T&H can be obtained and the imperfections (offset, gain-error and time-skew) can be determined.

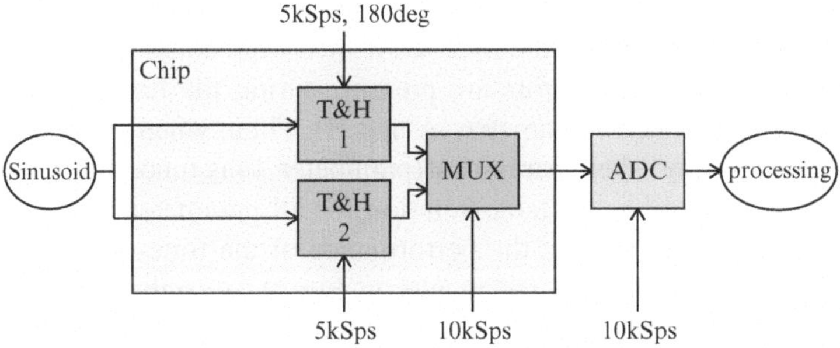

Figure 9.26. Setup for time-interleaved T&H performance verification.

Figure 9.27 shows the equipment used for both the self-measurement phase and the verification phase, which is identical to the setup described in Section 7. In this case, the FPGA board is also used to generate the MLS sequence and to control the multiplexer at the outputs of the T&H's.

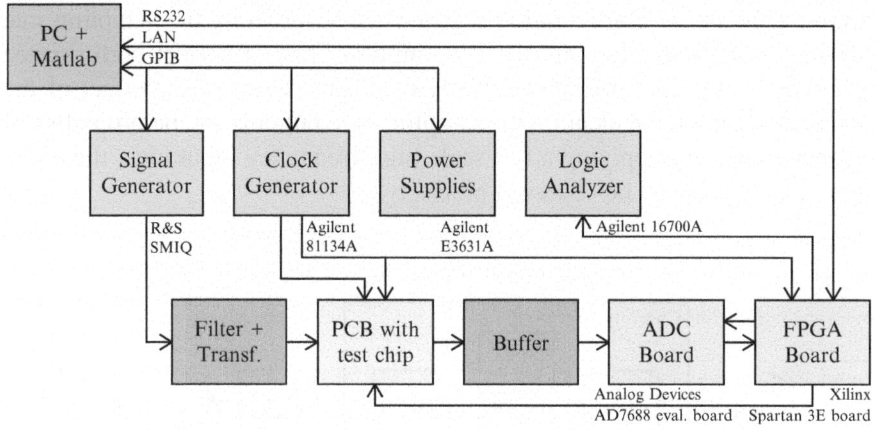

Figure 9.27. Time-interleaved T&H measurement setup.

8.2 Measurement results

During the measurements, a full-scale 238 MHz tone is applied as an input signal to the time-interleaved T&H. The output spectrum before calibration is shown in Fig. 9.28 (left). Three tones are indicated in the figure, the input tone (after downsampling), the spurious tone due to offset, and the spurious tone due to gain-error and time-skew. Both these spurious components limit the SNDR to 40 dB, or an ENOB of 6.3 bit. From the output data, the individual mismatches can be derived, yielding and offset of -4 mV, a gain-error of 1.25% and a time-skew of 11.3 ps.

Next, the calibration method was applied as described previously. Figure 9.29 shows the development of the three parameter settings as a function of the iteration of the algorithm. As the time-skew measurement is not precise when substantial gain and offset errors are present, during the first iterations only gain and offset are tuned (parameters A and B). Then, when a stable solution is found, also the time-skew parameter (parameter T) is tuned to the optimum value. After six iterations, a stable solution for all parameters is found. With these post-calibration settings, the performance of the time-interleaved T&H was verified again, obtaining the results in Fig. 9.28 (right). Both spurious tones are reduced by at least 15dB, resulting in a post-calibration ENOB of 8.9 bit. The post-calibration mismatches are as follows: an offset of -0.6 mV, a gain-error of 0.34% and a time-skew of -0.4 ps.

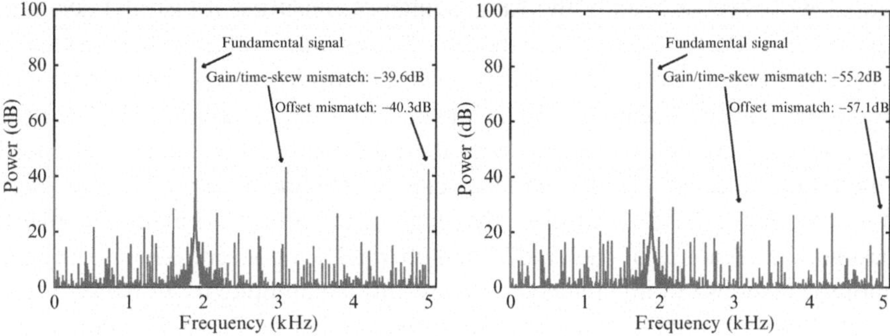

Figure 9.28. Measured output power spectrum (subsampled) for the 2-channel time-interleaved T&H, before calibration (left) and after calibration (right).

The post-calibration performance is around 1bit worse than expected. A possible reason is that during calibration and during verification, a slightly different setup is used. During calibration (Fig. 9.25), the inputs of the T&H's are connected together inside the chip, and then connected to the DAC. During verification (Fig. 9.26), the inputs of the T&H's are connected together outside the chip, and then connected to the signal source. As a result, during verification there is additional channel mismatch because of the separate traces on the PCB and the separate cables, which is not taken into account by the calibration method. For an improved implementation, the T&H inputs should be connected on-chip during the verification mode as well.

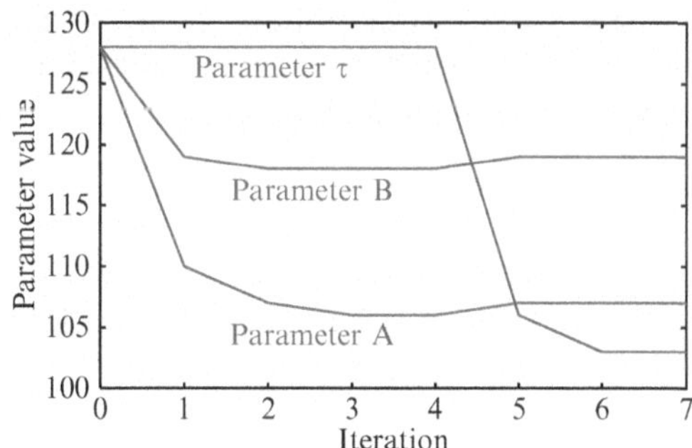

Figure 9.29. Parameter development as a function of the iteration of the algorithm.

9. Conclusion

In this chapter, a method was presented to measure and correct for offset, gain and time-skew errors in time-interleaved ADCs using a single integrated algorithm. The applied test signal is a repetitive pseudo-random noise sequence, which can be easily generated on-chip. Because of the repetitive nature, each channel will receive exactly the same sequence, ensuring a fast and accurate measurement. At the same time, the broad frequency range of the generated test signal prevents over-optimization for a specific frequency. Furthermore, it was shown that the accuracy and the speed of the DAC, used to generate the analog test signal, can be much lower than the intended accuracy and speed of the ADC. Because of that, the DAC can be integrated easily on chip without too much overhead. A further advantage of the presented method is that it measures the response of each individual channel as opposed to measuring the combined response of all channels. As a result, the complexity of the error detection algorithm remains proportional to the number of channels. Mathematical analysis proves that the resolution of the ADC is not limiting the accuracy of the measurement of the channel mismatches. Moreover, the measurement accuracy can be improved by increasing the length of the test-vector. As such, after calibration the overall performance will not be limited by the channel mismatches but by the performance of the individual channels. The actual correction is performed in the analog domain in such a way that little additional power is consumed. Chip measurements on a 2-channel time-interleaved T&H confirm that the calibration method can measure the channel mismatches and correct for them. In this specific implementation, the improvement is limited to 15 dB due to external imperfections, yielding a 9-bit post-calibration performance. However, this is not a general limitation as the analysis and simulation results prove that the channel-mismatches can be measured and corrected beyond the performance of the individual channels.

Notes

1. Matching between various channels is not only useful in case of time-interleaving, but also in systems where various parallel channels are used that are supposed to be identical, e.g. in a parallel readout circuit for an image sensor.
2. The actual DAC implementation has a sub-binary radix. However, in this context (without correction of the radix) it can be considered as an imperfect binary DAC.
3. As the sequence is periodic (in this example with a period length of 63 samples) a shift of e.g. -1 corresponds to a shift of $+62$.
4. This signal frequency is chosen as it is relative prime to the sample frequency and it is a relatively high frequency, and thus sensitive to time-skew errors.

Chapter 10

CONCLUSIONS

While technology evolution is beneficial for digital designs, it can cause problems or performance limitations for analog circuits. The smart concept can improve the performance of analog circuits by using on-chip intelligence to detect imperfections and to correct for them.

In a first example, considering a 12-bit DAC design, it is shown that a new smart approach can reduce the DAC area by a factor of 10 compared to existing state-of-the-art. For intrinsic designs, the minimum core area is determined by a limitation due to accuracy requirements, because of which the area increases with a factor of 4 for each additional bit. The presented approach, using a sub-binary variable-radix, overcomes the intrinsic limitation, because of which the area increases with a factor of 2 only for each additional bit. Thus, for higher resolutions, the approach becomes more and more attractive. Experimental results based on an implemented sub-binary variable-radix DAC prove the feasibility of the self-measurement method and the pre-correction algorithm.

In a second example, high-speed ADCs are considered. Smart approaches exist to correct for specific imperfections in ADCs. Even though they prove that these imperfections can be compensated, their overall performance (in terms of speed/accuracy/power) is not provably better compared to intrinsic designs. A limitation of most of the existing smart solutions is that their digital calibration hardware contributes significantly to the overall power consumption. In this work, start-up calibration methods are used in combination with analog correction circuits to maintain overall power-efficiency. Experimental results prove that analog correction of various imperfections is achievable in a power-efficient way.

In this work, two key-factors are perceived that enable high-performance: open-loop circuitry and time-interleaving. It is shown that open-loop circuits are able to achieve higher speed and lower power-consumption compared

P. Harpe et al., *Smart AD and DA Conversion*,
Analog Circuits and Signal Processing, DOI 10.1007/978-90-481-9042-3_10,

to closed-loop alternatives. However, a drawback of open-loop circuits is their limited accuracy. Mathematical analysis, simulations and experiments verify that the limited accuracy of open-loop circuits can be overcome by smart calibration. As such, the smart approach widens the application range of high-speed, power-efficient open-loop circuits towards higher accuracies. It is illustrated that the second key-factor, time-interleaving, enables a higher absolute speed of operation as well as a higher power-efficiency for specific situations. However, time-interleaving also introduces channel mismatch errors, which limit the final accuracy. A smart solution is proposed that can correct for these channel mismatches. Analysis and simulations prove that these mismatches can be measured and corrected accurately enough to ensure that the final post-calibration performance is not limited by the channel mismatches anymore, but by the performance of the individual channels. Experimental results verify the feasibility of the proposed methods for self-measurement and self-correction.

References

[1] J. E. Lilienfeld, "Method and apparatus for controlling electric currents," US Patent 1,745,175, filed Oct. 8, 1926.

[2] J. Bardeen and W. H. Brattain, "Three-electrode circuit element utilizing semiconductive materials," US Patent 2,524,035, filed June 17, 1948.

[3] W. Shockley, "Circuit element utilizing semiconductive material," US Patent 2,569,347, filed June 26, 1948.

[4] "UMC (United Microelectronics Corporation) – 2009," *from* www.umc.com.

[5] A. van Roermund, H. Hegt, P. Harpe, G. Radulov, A. Zanikopoulos, K. Doris, and P. Quinn, "Smart AD and DA converters," in *Proc. IEEE ISCAS 2005*, Kobe, Japan, May 23–26, 2005.

[6] B. Murmann and B. E. Boser, "A 12-bit 75-MS/s pipelined ADC using open-loop residue amplification," *IEEE J. Solid-State Circuits*, vol. 38, no. 12, pp. 2040–2050, Dec. 2003.

[7] W. Yang, D. Kelly, I. Mehr, M. T. Sayuk, and L. Singer, "A 3-V 340-mW 14-b 75-MSample/s CMOS ADC with 85-dB SFDR at Nyquist input," *IEEE J. Solid-State Circuits*, vol. 36, no. 12, pp. 1931–1936, Dec. 2001.

[8] P. J. A. Harpe, J. M. de Meulmeester, J. A. Hegt, and A. H. M. van Roermund, "Novel digital pre-correction method for mismatch in DACs with built-in self-measurement," in *Proc. IEE ADDA 2005*, July 25–27, 2005, pp. 25–30.

[9] P. Harpe, J. de Meulmeester, H. Hegt, and A. van Roermund, "Reliable design of digital-to-analog converters using pre-correction and embedded self-test," in *Proc. IEEE* 11th *Int. Mixed-Signals Testing Workshop*, June 27–29, 2005, pp. 84–89.

[10] P. Harpe, J. de Meulmeester, H. Hegt, and A. van Roermund, "Digital pre-correction method for mismatch in DACs with built-in self-measurement," in *Proc. ProRISC 2005*, Veldhoven, The Netherlands, Nov. 17–18, 2005, pp. 209–216.

[11] G. I. Radulov, "Flexible and self-calibrating current-steering digital-to-analog converters: Analysis, classification and design," Ph.D. dissertation, Eindhoven Univ. of Technology, Jan. 2010.

[12] W. Schofield, D. Mercer, and L. S. Onge, "A 16b 400MS/s DAC with <-80dBc IMD to 300MHz and <-160dBm/Hz noise power spectral density," in *Proc. IEEE ISSCC 2003*, 2003.

[13] G. I. Radulov, P. J. Quinn, H. Hegt, and A. van Roermund, "An on-chip self-calibration method for current mismatch in D/A converters," in *Proc. ESSCIRC 2005*, Sept. 12–16, 2005, pp. 169–172.

[14] Y. Tang, H. Hegt, and A. van Roermund, "Predictive timing error calibration technique for RF current-steering DACs," in *Proc. IEEE ISCAS 2008*, May 18–21, 2008, pp. 228–231.

[15] A. R. Bugeja, B.-S. Song, P. L. Rakers, and S. F. Gillig, "A 14-b, 100-MS/s CMOS DAC designed for spectral performance," *IEEE J. Solid-State Circuits*, vol. 34, no. 12, pp. 1719–1732, Dec. 1999.

[16] A. V. den Bosch, M. A. F. Borremans, M. S. J. Steyaert, and W. Sansen, "A 10-bit 1-GSample/s Nyquist current-steering CMOS D/A converter," *IEEE J. Solid-State Circuits*, vol. 36, no. 3, pp. 315–324, Mar. 2001.

[17] J. Deveugele and M. S. J. Steyaert, "A 10-bit 250-MS/s binary-weighted current-steering DAC," *IEEE J. Solid-State Circuits*, vol. 41, no. 2, pp. 320–329, Feb. 2006.

[18] K. Doris, "High-speed D/A converters: from analysis and synthesis concepts to IC implementation," Ph.D. dissertation, Eindhoven Univ. of Technology, Sept. 2004.

[19] K. O'Sullivan, C. Gorman, M. Hennessy, and V. Callaghan, "A 12-bit 320-MSample/s current-steering CMOS D/A converter in 0.44 mm^2," *IEEE J. Solid-State Circuits*, vol. 39, no. 7, pp. 1064–1072, July 2004.

[20] A. V. den Bosch, M. A. F. Borremans, M. S. J. Steyaert, and W. Sansen, "A 12-bit 500-MSample/s current-steering CMOS D/A converter," in *Proc. IEEE ISSCC 2001*, 2001.

[21] C.-H. Lin, F. van der Goes, J. Westra, J. Mulder, Y. Lin, E. Arslan, E. Ayranci, X. Liu, and K. Bult, "A 12b 2.9GS/s DAC with IM3 $\ll$-60dBc beyond 1GHz in 65nm CMOS," in *Proc. IEEE ISSCC 2009*, 2009, pp. 74–75.

[22] B. Schafferer and R. Adams, "A 3V CMOS 400mW 14b 1.4GS/s DAC for multi-carrier applications," in *Proc. IEEE ISSCC 2004*, 2004, pp. 360–532.

[23] J. Bastos, "Characterization of MOS transistor mismatch for analog design," Ph.D. dissertation, Katholieke Universiteit Leuven, 1998.

[24] T. Chen and G. G. E. Gielen, "A 14-bit 200-MHz current-steering DAC with switching-sequence post-adjustment calibration," *IEEE J. Solid-State Circuits*, vol. 42, no. 11, pp. 2386–2394, Nov. 2007.

[25] G. I. Radulov, P. J. Quinn, J. A. Hegt, and A. H. M. van Roermund, "A flexible 12-bit self-calibrated quad-core current-steering DAC," in *Proc. IEEE APPCAS 2008*, Nov. 30, 2008, pp. 25–28.

[26] Q. Huang, P. A. Francese, C. Martelli, and J. Nielsen, "A 200MS/s 14b 97mW DAC in 0.18µm CMOS," in *Proc. IEEE ISSCC 2004*, 2004.

[27] M. P. Tiilikainen, "A 14-bit 1.8-V 20-mW 1-mm^2 CMOS DAC," *IEEE J. Solid-State Circuits*, vol. 36, no. 7, pp. 1144–1147, July 2001.

[28] Y. Cong and R. L. Geiger, "A 1.5V 14b 100MS/s self-calibrated DAC," in *Proc. IEEE ISSCC 2003*, vol. 1, 2003, pp. 128–482.

[29] A. R. Bugeja and B.-S. Song, "A self-trimming 14b 100MSample/s CMOS DAC," in *Proc. IEEE ISSCC 2000*, 2000.

[30] P. G. A. Jespers, *Integrated Converters*. Oxford: Oxford University Press, 2001, ISBN 0 19 856446 5.

[31] M. Pastre and M. Kayal, "High-precision DAC based on a self-calibrated sub-binary radix converter," in *Proc. IEEE ISCAS 2004*, May 23–26, 2004, pp. 341–344.

[32] J. J. Wikner, "Studies on CMOS digital-to-analog converters," Ph.D. dissertation, Linkōpings Universitet, Sweden, 2001.

[33] A. Varzaghani and C.-K. K. Yang, "A 600-MS/s 5-bit pipeline A/D converter using digital reference calibration," *IEEE J. Solid-State Circuits*, vol. 41, no. 2, pp. 310–319, Feb. 2006.

[34] K. Iizuka, H. Matsui, M. Ueda, and M. Daito, "A 14-bit digitally self-calibrated pipelined ADC with adaptive bias optimization for arbitrary speeds up to 40 MS/s," *IEEE J. Solid-State Circuits*, vol. 41, no. 4, pp. 883–890, Apr. 2006.

[35] M. Daito, H. Matsui, M. Ueda, and K. Iizuka, "A 14-bit 20-MS/s pipelined ADC with digital distortion calibration," *IEEE J. Solid-State Circuits*, vol. 41, no. 11, pp. 2417–2423, Nov. 2006.

[36] B. Xia, A. Valdes-Garcia, and E. Sanchez-Sinencio, "A 10-bit 44-MS/s 20-mW configurable time-interleaved pipeline ADC for a dual-mode 802.11b/bluetooth receiver," *IEEE J. Solid-State Circuits*, vol. 41, no. 3, pp. 530–539, Mar. 2006.

[37] S. Ray and B.-S. Song, "A 13-b linear, 40-MS/s pipelined ADC with self-configured capacitor matching," *IEEE J. Solid-State Circuits*, vol. 42, no. 3, pp. 463–474, Mar. 2007.

[38] E. Iroaga and B. Murmann, "A 12-bit 75-MS/s pipelined ADC using incomplete settling," *IEEE J. Solid-State Circuits*, vol. 42, no. 4, pp. 748–756, Apr. 2007.

[39] Z.-M. Lee, C.-Y. Wang, and J.-T. Wu, "A CMOS 15-bit 125-MS/s time-interleaved ADC with digital background calibration," *IEEE J. Solid-State Circuits*, vol. 42, no. 10, pp. 2149–2160, Oct. 2007.

[40] Y.-S. Shu and B.-S. Song, "A 15-bit linear 20-MS/s pipelined ADC digitally calibrated with signal-dependent dithering," *IEEE J. Solid-State Circuits*, vol. 43, no. 2, pp. 342–350, Feb. 2008.

[41] I. Ahmed and D. A. Johns, "An 11-bit 45 MS/s pipelined ADC with rapid calibration of DAC errors in a multibit pipeline stage," *IEEE J. Solid-State Circuits*, vol. 43, no. 7, pp. 1626–1637, July 2008.

[42] J. Yuan, N. H. Farhat, and J. V. der Spiegel, "Background calibration with piecewise linearized error model for CMOS pipeline A/D converter," *IEEE Trans. Circuits Syst. I*, vol. 55, no. 1, pp. 311–321, Feb. 2008.

[43] B. P. Ginsburg and A. P. Chandrakasan, "Highly interleaved 5-bit, 250-MSamples/s, 1.2-mW ADC with redundant channels in 65-nm CMOS," *IEEE J. Solid-State Circuits*, vol. 43, no. 12, pp. 2641–2650, Dec. 2008.

[44] K.-W. Hsueh, Y.-K. Chou, Y.-H. Tu, Y.-F. Chen, Y.-L. Yang, and H.-S. Li, "A 1V 11b 200MS/s pipelined ADC with digital background calibration in 65nm CMOS," in *Proc. IEEE ISSCC 2008*, 2008, pp. 546–634.

[45] J. Hu, N. Dolev, and B. Murmann, "A 9.4-bit, 50-MS/s, 1.44-mW pipelined ADC using dynamic source follower residue amplification," *IEEE J. Solid-State Circuits*, vol. 44, no. 4, pp. 1057–1066, Apr. 2009.

[46] H. V. de Vel, B. A. J. Buter, H. van der Ploeg, M. Vertregt, G. J. G. M. Geelen, and E. J. F. Paulus, "A 1.2-V 250-mW 14-b 100-MS/s digitally calibrated pipeline ADC in 90-nm CMOS," *IEEE J. Solid-State Circuits*, vol. 44, no. 4, pp. 1047–1056, Apr. 2009.

[47] I. Ahmed, J. Mulder, and D. A. Johns, "A 50MS/s 9.9mW pipelined ADC with 58dB SNDR in 0.18μm CMOS using capacitive charge-pumps," in *Proc. IEEE ISSCC 2009*, 2009, pp. 164–165.

[48] E. Alpman, H. Lakdawala, L. R. Carley, and K. Soumyanath, "A 1.1V 50mW 2.5GS/s 7b time-interleaved C-2C SAR ADC in 45nm LP digital CMOS," in *Proc. IEEE ISSCC 2009*, 2009, pp. 76–77.

[49] S. Devarajan, L. Singer, D. Kelly, S. Decker, A. Kamath, and P. Wilkins, "A 16b 125MS/s 385mW 78.7dB SNR CMOS pipeline ADC," in *Proc. IEEE ISSCC 2009*, 2009, pp. 86–87.

[50] W. Liu, Y. Chang, S.-K. Hsien, B.-W. Chen, Y.-P. Lee, W.-T. Chen, T.-Y. Yang, G.-K. Ma, and Y. Chiu, "A 600MS/s 30mW 0.13μm CMOS ADC array achieving over 60dB SFDR with adaptive digital equalization," in *Proc. IEEE ISSCC 2009*, 2009, pp. 82–83.

[51] A. Panigada and I. Galton, "A 130mW 100MS/s pipelined ADC with 69dB SNDR enabled by digital harmonic distortion correction," in *Proc. IEEE ISSCC 2009*, 2009, pp. 162–163.

[52] R. C. Taft, P. A. Francese, M. R. Tursi, O. Hidri, A. Mackenzie, T. Hoehn, P. Schmitz, H. Werker, and A. Glenny, "A 1.8V 1.0GS/s 10b self-calibrating unified-folding-interpolating ADC with 9.1 ENOB at Nyquist frequency," in *Proc. IEEE ISSCC 2009*, 2009, pp. 78–79.

[53] A. Verma and B. Razavi, "A 10b 500MHz 55mW CMOS ADC," in *Proc. IEEE ISSCC 2009*, 2009, pp. 84–85.

[54] S.-T. Ryu, B.-S. Song, and K. Bacrania, "A 10b 50MS/s pipelined ADC with opamp current reuse," in *Proc. IEEE ISSCC 2006*, 2006.

[55] B.-G. Lee, B.-M. Min, G. Manganaro, and J. W. Valvano, "A 14b 100MS/s pipelined ADC with a merged active S/H and first MDAC," in *Proc. IEEE ISSCC 2008*, 2008, pp. 248–611.

[56] M. Boulemnakher, E. Andre, J. Roux, and F. Paillardet, "A 1.2V 4.5mW 10b 100MS/s pipeline ADC in a 65nm CMOS," in *Proc. IEEE ISSCC 2008*, 2008, pp. 250–251.

[57] B.-G. Lee and R. M. Tsang, "A 10-bit 50 MS/s pipelined ADC with capacitor-sharing and variable-g_m opamp," *IEEE J. Solid-State Circuits*, vol. 44, no. 3, pp. 883–890, Mar. 2009.

[58] B. Razavi, *Design of Analog CMOS Integrated Circuits*. New York: McGraw-Hill, 2001, ISBN 0-07-118815-0.

[59] W. C. Black and D. A. Hodges, "Time interleaved converter arrays," *IEEE J. Solid-State Circuits*, vol. 15, no. 12, pp. 1022–1029, Dec. 1980.

[60] C. Vogel and H. Johansson, "Time-interleaved analog-to-digital converters: Status and future directions," in *Proc. IEEE ISCAS 2006*, May 21–24, 2006, pp. 3386–3389.

[61] P. Harpe, A. Zanikopoulos, H. Hegt, and A. van Roermund, "A 62dB SFDR, 500MSPS, 15mW open-loop track-and-hold circuit," in *Proc.* 24th *Norchip 2006*, Linkoping, Sweden, Nov. 20–21, 2006, pp. 103–106.

[62] P. Harpe, H. Hegt, and A. van Roermund, "Analysis of open-loop track-and-hold circuits," in *Proc. IEEE ICECS 2007*, Dec. 11–14, 2007, pp. 1236–1239.

[63] S. M. Louwsma, E. J. M. van Tuijl, M. Vertregt, P. C. S. Scholtens, and B. Nauta, "A 1.6 GS/s, 16 times interleaved track&hold with 7.6 ENOB in 0.12μm CMOS," in *Proc. ESSCIRC 2004*, Sept. 2004, pp. 343–346.

[64] S. M. Louwsma, E. J. M. van Tuijl, M. Vertregt, and B. Nauta, "A time-interleaved track&hold in 0.13μm CMOS sub-sampling a 4 GHz signal with 43dB SNDR," in *Proc. IEEE CICC 2007*, Sept. 16–19, 2007, pp. 329–332.

[65] A. Baschirotto, "A low-voltage sample-and-hold circuit in standard CMOS technology operating at 40MS/s," *IEEE Trans. Circuits Syst. II*, vol. 48, pp. 394–399, Apr. 2001.

[66] M. Chennam and T. S. Fiez, "A 0.35μm current-mode T&H with -81dB THD," in *Proc. IEEE ISCAS 2004*, May 23–26, 2004, pp. 1112–1115.

[67] D. Vecchi, C. Azzolini, A. Boni, F. Chaahoub, and L. Crespi, "100-MS/s 14-b track-and-hold amplifier in 0.18-μm CMOS," in *Proc. ESSCIRC 2005*, Sept. 2005, pp. 259–262.

[68] S. Chatterjee and P. R. Kinget, "A 0.5-V 1-MSps track-and-hold circuit with 60-dB SNDR," *IEEE J. Solid-State Circuits*, vol. 42, pp. 722–729, Apr. 2007.

[69] T.-S. Lee, C.-C. Lu, and C.-C. Ho, "A 330MHz 11bit 26.4mW CMOS low-hold-pedestal fully differential track-and-hold circuit," in *Proc. IEEE VLSI-DAT Symposium 2008*, Apr. 23–25, 2008, pp. 144–147.

[70] A. Boni, A. Pierazzi, and C. Morandi, "A 10-b 185-MS/s track-and-hold in 0.35μm CMOS," *IEEE J. Solid-State Circuits*, vol. 36, pp. 195–203, Feb. 2001.

[71] D. Jakonis and C. Svensson, "A 1GHz linearized CMOS track-and-hold circuit," in *Proc. IEEE ISCAS 2002*, May 26–29, 2002, pp. 577–580.

[72] I. H. Wang, J. L. Lin, and S. I. Liu, "5-bit, 10 Gsamples/s track-and-hold circuit with input feedthrough cancellation," *Electronics Letters*, vol. 42, pp. 457–459, Apr. 2006.

[73] A. M. Abo and P. R. Gray, "A 1.5-V, 10-bit, 14.3-MS/s CMOS pipeline analog-to-digital converter," *IEEE J. Solid-State Circuits*, vol. 34, no. 5, pp. 599–606, May 1999.

[74] S. Ouzounov, E. Roza, J. A. Hegt, G. van der Weide, and A. H. M. van Roermund, "A CMOS V-I converter with 75-dB SFDR and 360-μW power consumption," *IEEE J. Solid-State Circuits*, vol. 40, no. 7, pp. 1527–1532, July 2005.

[75] P. Harpe, A. Zanikopoulos, H. Hegt, and A. van Roermund, "Analog calibration of mismatches in an open-loop track-and-hold circuit for time-interleaved ADCs," in *Proc. IEEE ISCAS 2007*, May 27–30, 2007, pp. 1951–1954.

[76] P. Harpe, A. Zanikopoulos, H. Hegt, and A. van Roermund, "Analog calibration of an open-loop track-and-hold circuit," in *Proc. ProRISC 2007*, Veldhoven, The Netherlands, Nov. 29–30, 2007.

[77] P. Harpe, H. Hegt, and A. van Roermund, "Analog calibration of channel mismatches in time-interleaved ADCs," in *Proc. IEEE ECCTD 2007*, Aug. 26–30, 2007, pp. 236–239.

[78] P. J. A. Harpe, J. A. Hegt, and A. H. M. van Roermund, "Analog calibration of channel mismatches in time-interleaved ADCs," *Int. J. Circ. Theor. Appl.*, vol. 37, no. 2, pp. 301–318, Mar. 2009.

[79] K. C. Dyer et al., "An analog background calibration technique for time-interleaved analog-to-digital converters," *IEEE J. Solid-State Circuits*, vol. 33, no. 12, pp. 1912–1919, Dec. 1998.

[80] H. Johansson and P. Löwenborg, "Reconstruction of nonuniformly sampled bandlimited signals by means of digital fractional delay filters," *IEEE Trans. Signal Processing*, vol. 50, no. 11, pp. 2757–2767, Nov. 2002.

[81] S. M. Jamal et al., "Calibration of sample-time error in a two-channel time-interleaved analog-to-digital converter," *IEEE Trans. Circuits Syst. I*, vol. 51, no. 1, pp. 130–139, Jan. 2004.

[82] C. Vogel, D. Draxelmayr, and F. Kuttner, "Compensation of timing mismatches in time-interleaved analog-to-digital converters through transfer characteristics tuning," in *Proc. IEEE MWSCAS 2004*, 2004, pp. 341–344.

[83] C. Vogel, D. Draxelmayr, and G. Kubin, "Spectral shaping of timing mismatches in time-interleaved analog-to-digital converters," in *Proc. IEEE ISCAS 2005*, 2005, pp. 1394–1397.

[84] S. W. Golomb, *Shift Register Sequences*. Holden-Day, Inc., 1967.

Index

A
AD and DA conversion
general view, 3–4
performance criteria, 8
trends in applications, 4–5
trends in system design
design time and risk, 7
flexibility, 7
hostile environment, 6
resource reuse, 7–8
stand-alone analog component testing, 6
system optimization, 8
technology portability, 7
yield, integrated components, 6–7
trends in technology
C-value, 5–6
higher speed, 5
lower power-consumption, 5
low voltage operation, 6
short channel effects, 6
smaller area, 5
technology and power supply scaling, 5
Analog components, 2
Analog correction circuits, 155

B
Binary architecture, 26

C
Cadence simulation
analog correction parameters, 111
sub-binary variable-radix DAC, 53–54
Channel mismatch
calibration
system overview, 129
test-signal generation, 129–131
correction
analog correction method, 145–147
correction algorithm, 147
detection
channel responses, 4-channel ADC, 132, 133
frequency spectrum and autocorrelation function, 131, 133
gain error detection, 134–136
iterative procedure, 133
offset detection, 133–134
time-skew detection, 136–145
model, 131–132
Current-source-array (CSA)
current-steering DACs, 21–22
sub-binary variable-radix DAC, 44–45

D
Differential non-linearity (DNL), 26–27, 37, 53, 54
Differential pair, 79
Digital calibration hardware, 155
Digitally assisted analog correction
optimization algorithm
actual data points *vs.* linear estimation, 114
distortion estimation, 116
error estimation, 114, 116
non-linearity measurement, 115
self-measurement method, 113–114
DNL. *See* Differential non-linearity

E
Effective-number-of-bits (ENOB), 73, 74, 77, 84, 98, 99
Effective resolution bandwidth (ERBW), 58
Enhanced differential pair
cross-coupling, 85
ENOB, 84
gain and fifth-order distortion, 86–87

maximum deviation, 83, 84
output buffer parameters, 87, 88
output voltage, V_{out}, 82, 83, 85
polynomial model, 86
resistive load, 82–84
resistive source degeneration, 84, 85
simulated distortion, output buffer, 87, 88
simulated noise power, output buffer, 87, 89
third-order distortion, 83
transfer function, 85–86
unity-gain, 83

ENOB. *See* Effective-number-of-bits
Environmental information, 13
ERBW. *See* Effective resolution bandwidth

F

Figure-of-merit (FoM), 57–60, 74
FPGA, 121, 122

G

Gain-bandwidth-product (GBP), 62

H

High-speed high-resolution AD conversion
open-loop *vs.* closed-loop circuitry, 64
amplifiers, 61–62
feedback, 62
power consumption, 63
small-signal model, 62
smart properties, 60
time-interleaving
2-channel solution, 66
parallelism, 67
pole frequency, 64–65
power and speed trade-off, 66
sampling rate, 64
source follower, 64, 65
speed scales, 65
T&H circuit, 64

I

Integral non-linearity (INL), 37, 53, 54
Integrated circuits (ICs), 1
Interference compatibility, 16

K

Kilby, J., 1

L

Latch, 42
Lilienfeld, Julius, 1

M

Maximum-length sequence (MLS), 129–130
Mentor Eldo simulator, 99
Mixed-signal ICs, 1
Mixed-signal integration, 1–2
MLS. *See* Maximum-length sequence
Monte-Carlo simulation, 37, 110

N

Noyce, R., 1
Nyquist band, 91

O

On-chip test-signal generator, 20
Open-loop circuitry, 155–156
Open-loop track and hold (T&H) circuit design
vs. closed-loop solutions, 74
CMOS technology, 73–74
design specifications, 75
differential pair based architecture, 89
ENOB, 73, 74
feasibility, 75
features, 91
floorplan and layout, 91, 93
FoM, 74
gain *vs.* input frequency, 90
linearity and noise, 73
measurement results
jitter performance, 99
performance, 100–102
sampling core, 100, 101
setup, 95, 97
SFDR *vs.* input frequency, 97, 98
transient noise power and sinusoidal input signal, 99, 101
measurement setup
input and output signals, 91–93
off-chip ADC, 93–95
PCB, 95–96
subsampling, 94, 95
output buffer architecture
accuracy, mismatch sensitivity and controllability, 79–81
differential pair, 79
enhanced differential pair, 82–89
power consumption and speed, 76, 79, 80
power supply requirements and portability, 79, 81–82
source follower, 78–79
sampling core architecture
bootstrapping technique, 77
distortion curve, 77, 78
sampling capacitors, switches and switch drivers, 76–77
transistor-level simulations, 77
SFDR, 90, 92
simulated configurations, 89, 90
source follower based architecture, 79, 89
speed and linearity, 75, 76
stand-alone component, 75

static linearity, 90–92
T&H architecture, 75–76
transistor-level simulations, 89

P
Programmable clock delay circuit, 149, 150
Pseudo-differential source follower, 78–79

S
Sampling core architecture
bootstrapping technique, 77
distortion curve, 77, 78
sampling capacitors, switches and switch drivers, 76–77
transistor-level simulations, 77
Segmented architecture, 26
Serial clock, 41
Serial data, 42
Serial-in parallel-out register, 41–42
SFDR. *See* Spurious free dynamic range
Shockley, W., 1
Signal information, 13
Smart AD conversion
digital calibration method, 59
ERBW, 58
intrinsic and calibrated ADCs, 58, 61
mixed analog/digital system, 57
open-loop *vs.* closed-loop circuitry, 64
amplifiers, 61–62
feedback, 62
power consumption, 63
small-signal model, 62
pipelined AD converters, 59, 60
recent publications, 58, 59
smart calibration
analog control, 69, 70
channel mismatch, 67
coarse error correction, 69
digital register, 69
foreground and background calibration, 68, 69
smart conversion, 68
smart properties, 60
time-interleaving
2-channel solution, 66
parallelism, 67
pole frequency, 64–65
power and speed trade-off, 66
sampling rate, 64
source follower, 64, 65
speed scales, 65
T&H circuit, 64
Smart converter concept
analog correction, 15
application
flexibility, 16
portability, design time and risk, 16
speed, accuracy, power consumption, area, 15
testability and yield, 16
digital correction, 15
information extraction
environmental information, 13
optimization algorithm, 14
processing algorithm, 12, 14
signal information, 13
system information, 12–13
mapping method, 15
Smart DA conversion
analog and digital scaling, 21
binary-scaled design, 34
13-bit DAC design, 39
correction, calibration and mapping, 19
current-steering DACs
architecture, 21
CSA, 21–22
decoded digital signal, 21
intrinsic-design DA converter, 22
3σ mismatch-error, 22–23
total gate-area, 23
digital pre-correction algorithm, 33
error mechanism, 19
independent current sources, 38
INL and DNL, 37
large-scale mixed-signal ICs, 20–21
mismatch error correction, 24–25
Monte-Carlo simulations, 37
overall DAC and analog core area, 20
sub-binary fixed-radix design, 37, 40
sub-binary variable-radix design, 39
current sources, 35–36
full-scale current, 36
input code 827, 36, 37
recursive measurement algorithm, 35
residue, 36
standard deviation, 34
technology-scaling, 20
Spurious free dynamic range (SFDR), 90, 92, 104, 106, 122
Sub-binary variable-radix DAC
binary architecture, 26
built-in measurement algorithm, 25–26
CSA design, 44–45
digitally pre-corrected DAC, 25, 26
digital re-mapping, 27
DNL error, 26–27
implemented current-steering DAC, 41, 42
layout, 45–46
measurement results
vs. Cadence simulation, 53–54
cell approach, 51
INL and DNL, 53, 54
input mapping, 52

pre-correction method, 51–52
redundancy, 53
self-measurement, 50–51
sub-DACs, 52–53
systematic mismatch error, 51
uncorrected transfer curve, 53
measurement setup
digital pre-correction, transfer curve, 49–50
self-measurement, current sources, 47–48
transfer curve determination, 48–49
mismatch-error, 25
redundancy
error probability, 30
N-bit binary converter, 28
relative spread, 29
stochastic spread, 28
self-measurement
analog measurement circuit, 32–33
circuit implementation, 46–47
measurement algorithm, 30–32
serial-in parallel-out register, 41–42
simulation results, 45
switch driver, 42
switches, 42–43
transfer curve, 5-bit DAC, 26
System information, 12–13
System integration, 1

T

Test-signal generation
filter, 130
MLS, 129–130
model, shift register and DAC, 130
normalized power and normalized auto-correlation, 130, 131
on-chip signal generator, ADC, 129
Thermometer architecture, 26
Time-interleaved ADCs
accuracy, 125
calibration method and layout implementation, 149–150
channel matching, T&H, 128
channel mismatch
calibration, 129–131
channel responses, 4-channel ADC, 132, 133
correction, 145–147
frequency spectrum and autocorrelation function, 131, 133
gain error detection, 134–136
iterative procedure, 133
model, 131–132
offset detection, 133–134
time-skew detection, 136–145
disadvantages, 127
matching errors, 126
measurement results, 152–153
measurement setup
self-measurement phase, 150–152
T&H performance verification, 151
verification phase, 152
mismatch errors, 127
N-bit p-channel, 125, 126
simulation results
HD3 component, 148–149
offset, gain-error and time-skew *vs.* temperature, 149
output spectrum, 147, 148
self-measurement and self-correction algorithms, 148
stability, 149
transistor-level simulations, 147
Time-interleaving, 155–156
Time-skew, 132
Time-skew detection
cross-correlation
difference and time-skew error relation, 138
discrete cross-correlation, 138, 139
$D_2[p]$ and estimated error, 140
frequency spectrum and autocorrelation function, 131, 139
standard deviation, *Di*[p], 140
time-skew, Δ_i, 139, 140
frequency-domain analysis, Z-domain
frequency response, 141, 143
measured response, $U_i[k]$, 141, 142
output of channel, 141
time-skew estimation, 142, 143
multi-bit MLS sequence length, 136–137
time-interleaved converter, 137, 138
time-skew estimation, 142, 143
accuracy, 144–145
methods, comparison, 144
Track and hold (T&H) calibration. *See also* Time-interleaved ADCs
analog correction parameters
Cadence simulation, 111
controllable current sources, 108
design parameters, 108–109
differential pair with resistive load, 106, 107
digital control signal, 108
enhanced differential pair with programmable tail current sources, 107
even- and odd-order distortion, 106, 112–113
main and cross-coupled pair, 110
Monte-Carlo simulation, 110
offset and gain controllability, 112
post-calibration performance, 111–112

SFDR metric, 106
simplified single-ended view, 109
transistor-level design, 111
behavioral-level simulations, 117–118
calibration method, 105–106
gain error and offset optimization, 104
implementation, 119–120
measurement results
linearity optimization, 123–124
measured offset, 122, 123
relative gain-error, 122, 123
transfer curve, 122
measurement setup
analog test signal, 121
FPGA, 121, 122
mixed-signal board, 121
ramp signal, 122
self-measurement phase, 120, 121
verification phase, 121
optimization algorithm
actual data points *vs.* linear estimation, 114
distortion estimation, 116
error estimation, 114, 116
non-linearity measurement, 115
self-measurement method, 113–114
sub-optimal accuracy, 103
T&H accuracy, 104–105
transistor-level simulations, 119

The manufacturer's authorised representative in the EU is Springer Nature Customer Service Centre GmbH, Europaplatz 3, 69115 Heidelberg, Germany. If you have any concerns regarding our products, please contact ProductSafety@springernature.com

Printed and bound by CPI Group (UK) Ltd, Croydon, CR0 4YY
15/07/2026
02167656-0002